Tucholsky Wagner Zola Scott Sydow Freud Schlegel
Turgenev Fonatne Wallace
Twain Walther von der Vogelweide Fouqué Friedrich II. von Preußen
Weber Freiligrath Frey
Kant Ernst
Fechner Fichte Weiße Rose von Fallersleben Richthofen Frommel
Hölderlin
Engels Fielding Eichendorff Tacitus Dumas
Fehrs Faber Flaubert
Maximilian I. von Habsburg Fock Eliasberg Ebner Eschenbach
Feuerbach Zweig
Eliot Vergil
Ewald
Goethe London
Elisabeth von Österreich
Mendelssohn Balzac Shakespeare Dostojewski Ganghofer
Lichtenberg Rathenau Doyle Gjellerup
Trackl Stevenson Hambruch
Mommsen Tolstoi Lenz Droste-Hülshoff
Thoma Hanrieder
von Arnim Hägele Hauff Humboldt
Dach Verne
Reuter Rousseau Hagen Hauptmann Gautier
Karrillon Garschin
Defoe Baudelaire
Damaschke Hebbel
Descartes
Hegel Kussmaul Herder
Wolfram von Eschenbach Dickens Schopenhauer
Darwin Grimm Jerome Rilke George
Bronner Melville
Bebel Proust
Campe Horváth Aristoteles
Barlach Voltaire Federer Herodot
Bismarck Vigny Heine
Gengenbach
Storm Casanova Tersteegen Grillparzer Georgy
Gilm
Chamberlain Lessing Langbein Gryphius
Brentano Lafontaine
Strachwitz Claudius Schiller Kralik Iffland Sokrates
Bellamy Schilling
Katharina II. von Rußland Gerstäcker Raabe Gibbon Tschechow
Löns Hesse Hoffmann Gogol Wilde Vulpius
Gleim
Luther Heym Hofmannsthal Morgenstern
Klee Hölty Goedicke
Roth Heyse Klopstock Kleist
Luxemburg Puschkin Homer Mörike
La Roche Horaz Musil
Machiavelli Kierkegaard Kraft Kraus
Navarra Aurel Musset Moltke
Lamprecht Kind Kirchhoff Hugo
Nestroy Marie de France
Laotse Ipsen Liebknecht
Nietzsche Nansen
Marx Ringelnatz
von Ossietzky Lassalle Gorki Klett Leibniz
May
vom Stein Lawrence Irving
Petalozzi Knigge
Platon
Pückler Michelangelo Kafka
Sachs Poe Kock
Liebermann
de Sade Praetorius Mistral Zetkin Korolenko

The publishing house tredition has created the series **TREDITION CLASSICS**. It contains classical literature works from over two thousand years. Most of these titles have been out of print and off the bookstore shelves for decades.

The book series is intended to preserve the cultural legacy and to promote the timeless works of classical literature. As a reader of a **TREDITION CLASSICS** book, the reader supports the mission to save many of the amazing works of world literature from oblivion.

The symbol of **TREDITION CLASSICS** is Johannes Gutenberg (1400 – 1468), the inventor of movable type printing.

With the series, tredition intends to make thousands of international literature classics available in printed format again – worldwide.

All books are available at book retailers worldwide in paperback and in hardcover. For more information please visit: www.tredition.com

tredition was established in 2006 by Sandra Latusseck and Soenke Schulz. Based in Hamburg, Germany, tredition offers publishing solutions to authors and publishing houses, combined with worldwide distribution of printed and digital book content. tredition is uniquely positioned to enable authors and publishing houses to create books on their own terms and without conventional manufacturing risks.

For more information please visit: www.tredition.com

The Dyeing of Cotton Fabrics A Practical Handbook for the Dyer and Student

Franklin Beech

Imprint

This book is part of the TREDITION CLASSICS series.

Author: Franklin Beech
Cover design: toepferschumann, Berlin (Germany)

Publisher: tredition GmbH, Hamburg (Germany)
ISBN: 978-3-8491-9227-3

www.tredition.com
www.tredition.de

PREFACE.

In writing this little book the author believes he is supplying a want which most Students and Dyers of Cotton Fabrics have felt — that of a small handbook clearly describing the various processes and operations of the great industry of dyeing Cotton.

The aim has not been to produce a very elaborate treatise but rather a book of a convenient size, and in order to do so it has been necessary to be brief and to omit many matters that would rightfully find a place in a larger treatise, but the author hopes that nothing of importance has been omitted. The most modern processes have been described in some detail; care has been taken to select those which experience shows to be thoroughly reliable and to give good results.

FRANKLIN BEECH.

May, 1901.

CONTENTS.

[Pg vi]

CHAPTER I.

STRUCTURE AND CHEMISTRY OF THE COTTON FIBRE.

There is scarcely any subject of so much importance to the bleacher, textile colourist or textile manufacturer as the structure and chemistry of the cotton fibre with which he has to deal. By the term chemistry we mean not only the composition of the fibre substance itself, but also the reactions it is capable of undergoing when brought into contact with various chemical substances—acids, alkalies, salts, etc. These reactions have a very important bearing on the operations of bleaching and dyeing of cotton fabrics.

A few words on vegetable textile fibres in general may be of interest. Fibres are met with in connection with plants in three ways.

First, as cuticle or ciliary fibres or hairs; these are of no practical use, being much too short for preparing textile fabrics from, but they play an important part in the physiology of the plant.

Second, as seed hairs; that is fibres that are attached to the seeds of many plants, such, for instance, as the common thistle and dandelion; the cotton fibre belongs to this group of seed hairs, while there are others, kapok, etc., that have been tried from time to time in spinning and weaving, but without much success. These seed hairs vary much in length, from ¼ inch to 1½ inches or even 2 inches; each fibre consists of a single unit. Whether it is serviceable as a textile fibre [Pg 2] depends upon its structure, which differs in different plants, and also upon the quantity available.

The third class of fibre, which is by far the most numerous, consists of those found lying between the bark or outer cuticle and the true woody tissues of the plant. This portion is known as the bast, and hence these fibres are known as "bast fibres". They are noticeable on account of the great length of the fibres, in some cases upwards of 6 feet, which can be obtained; but it should be pointed out that these long fibres are not the unit fibres, but are really bundles of the ultimate fibres aggregated together to form one long fibre, as found in and obtained from the plant. Thus the ultimate fibres of jute are really very short—from 1/10 to 1/8 of an inch in length;

those of flax are somewhat longer. Jute, flax, China grass and hemp are common fibres which are derived from the bast of the plants.

There is an important point of difference between seed fibres and bast fibres, that is in the degree of purity. While the seed fibres are fairly free from impurities—cotton rarely containing more than 5 per cent.—the bast fibres contain a large proportion of impurity, from 25 to 30 per cent. as they are first obtained from the plant, and this large quantity has much influence on the extent and character of the treatments to which they are subjected.

As regards the structure of the fibres, it will be sufficient to say that while seed hairs are cylindrical and tubular and have thin walls, bast fibres are more or less polygonal in form and are not essentially tubular, having thick walls and small central canals.

The Cotton Fibre.—The seed hairs of the cotton plant are separated from the seeds by the process of ginning, and they then pass into commerce as raw cotton. In this condition the fibre is found to consist of the actual fibrous substance itself, containing, however, about 8 per cent. of hygroscopic [Pg 3] or natural moisture, and 5 per cent. of impurities of various kinds, which vary in amount and in kind in various descriptions of cotton. In the process of manufacture into cotton cloths, and as the material passes through the operations of bleaching, dyeing or printing, the impurities are eliminated.

Impurities of the Cotton Fibre.—Dr. E. Schunck made an investigation many years ago into the character of the impurities, and found them to consist of the following substances:—

Cotton Wax.—This substance bears a close resemblance to carnauba wax. It is lighter than water, has a waxy lustre, is somewhat translucent, is easily powdered, and melts below the boiling point of water. It is insoluble in water, but dissolves in alcohol and in ether. When boiled with weak caustic soda it melts but is not dissolved by the alkali; it can, however, be dissolved by boiling with alcoholic caustic potash. This wax is found fairly uniformly distributed over the surface of the cotton fibre, and it is due to this fact that raw cotton is wetted by water only with difficulty.

Fatty Acids. — A solid, fatty acid, melting at 55° C. is also present in cotton. Probably stearic acid is the main constituent of this fatty acid.

Colouring Matter. — Two brown colouring matters, both containing nitrogen, can be obtained from raw cotton. One of these is readily soluble in alcohol, the other only sparingly so. The presence in relatively large quantities of these bodies accounts for the brown colour of Egyptian and some other dark-coloured varieties of cotton.

Pectic Acid. — This is the chief impurity found in raw cotton. It can be obtained in the form of an amorphous substance of a light yellow colour, not unlike gum in appearance. It is soluble in boiling water, and the solution has a faint acid reaction. Acids and many metallic [Pg 4] salts, such as mercury, chloride and lead acetate, precipitate pectic acid from its solutions. Alkalies combine with it, and these compounds form brown substances, are but sparingly soluble in water, and many of them can be precipitated out by addition of neutral salts, like sodium and ammonium chlorides.

Albumens. — A small quantity of albuminous matter is found among the impurities of cotton.

Structure of the Cotton Fibre. — The cotton fibre varies in length from 1 to 2 inches, not only in fibres of the same class but also in fibres from different localities — Indian fibres varying from 0.8 in the shortest to 1.4 in the longest stapled varieties; Egyptian cotton fibres range from 1.1 to 1.6 inches long; American cotton ranges from 0.8 in the shortest to 2 inches in the longest fibres. The diameter is about 1/1260 of an inch. When seen under the microscope fully ripe cotton presents the appearance of irregularly twisted ribbons, with thick rounded edges. The thickest part is the root end, or point of attachment to the seed. The free end terminates in a point. The diameter is fairly uniform through ¾ to ⅞ of its length, the rest is taper. In Fig. 1 is given some illustrations of the cotton fibre, showing this twisted and ribbon-like structure, while in Fig. 1A is given some transverse sections of the fibre. These show that it is a collapsed cylinder, the walls being of considerable thickness when compared with the internal bore or canal.

Perfectly developed, well-formed cotton fibres always present this appearance. But all commercial cottons contain more or less of fibres which are not perfectly developed or are unripe. These are known as "dead fibres"; they do not spin well and they do not dye well. On examination under the microscope it is seen that these fibres have not the flattened, twisted appearance of the ripe fibres, but are flatter, [Pg 5] and the central canal is almost obliterated and the fibres are but little twisted. Dead fibres are thin, brittle and weak.

Composition of the Cotton Fibres.—Of all the vegetable textile fibres cotton is found to have the simplest chemical composition and to be, as it were, the type substance of all such fibres, the others differing from it in several respects. When stripped of the comparatively small quantities of impurities, cotton is found to consist of a substance to which the name of cellulose has been given.

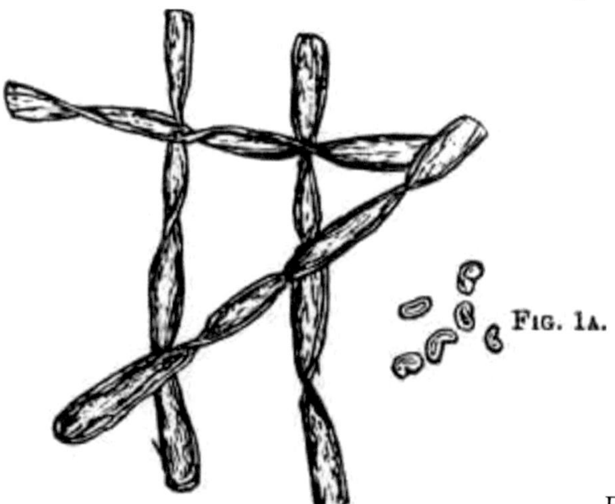

FIG. 1.—
Cotton Fibre.

Cellulose is a compound of the three elements, carbon, hydrogen and oxygen, in the proportions shown in the following analysis:—

Carbon, 44.2 per cent., Hydrogen, 6.3 per cent., Oxygen, 49.5 per cent., which corresponds to the empirical formula $C_6H_{10}O_5$, which shows it to belong to the group of carbo-hydrates, that is, bodies which contain the hydrogen and oxygen present in them in the proportion in [Pg 6] which they are present in water, namely H_2O.

Cellulose may be obtained in a pure condition from cotton by treatment with alkalies, followed by washing, and by treatment with alkaline hypochlorites, acids, washing and, finally, drying. As thus obtained it is a white substance having the form of the fibre from which it is procured, showing a slight lustre, and is slightly translucent. The specific gravity is 1.5, it being heavier than water. It is characterised by being very inert, a property of considerable value from a technical point of view, as enabling the fibres to stand the various operations of bleaching, dyeing, printing, finishing, etc. Nevertheless, by suitable means, cellulose can be made to undergo various chemical decompositions which will be noted in some detail.

Cellulose on exposure to the air will absorb moisture or water. This is known as hygroscopic moisture, or "water of condition". The amount in cotton is about 8 per cent., and it has a very important bearing on the spinning properties of the fibre, as it makes the fibre soft and elastic, while absolutely dry cotton fibre is stiff, brittle and non-elastic; hence it is easier to spin and weave cotton in moist climates or weather than in dry climates or weather. Cotton cellulose is insoluble in all ordinary solvents, such as water, ether, alcohol, chloroform, benzene, etc., and these agents have no influence in any way on the material, but it is soluble in some special solvents that will be noted later on.

ACTION OF ALKALIES.

The action of alkalies on cellulose or cotton is one of great importance in view of the universal use of alkaline liquors made from soda or caustic soda in the scouring, bleaching and dyeing of cotton, while great interest attaches to the use of caustic soda in the "mercerising" of cotton.

Dilute solutions of the caustic alkalies, caustic soda or caustic potash, of from 2 to 7 per cent. strength, have no [Pg 7] action on cellulose or cotton, in the cold, even when a prolonged digestion of the fibre with the alkaline solution takes place. Caustic alkali solutions of from 1 to 2 per cent. strength have little or no action even when used at high temperatures and under considerable pressure— a fact of very great importance from a bleacher's point of view, as it enables him to subject cotton to a boil in kiers, with such alkaline solutions at high pressures, for the purpose of scouring the cotton, without damaging the fibre itself.

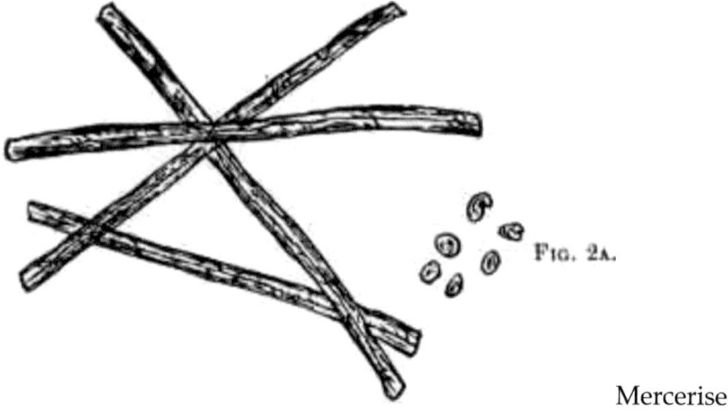

Mercerised Cotton Fibre.

Solutions of caustic soda of greater strength than 3 per cent. tend, when boiled under pressure, to convert the cellulose into soluble bodies, and as much as 20 per cent. of the fibre may become dissolved under such treatment. The action of strong solutions of caustic soda or caustic potash upon cellulose or cotton is somewhat different. Mercer found that solutions containing 10 per cent. of alkali had a very considerable effect upon the fibre, causing it to swell up and become gelatinous and transparent in its structure, each individual cotton fibre losing its ribbon-like appearance, and assuming a rod-like form, the central canal being more or less obliterated. This is shown in Fig. 2 and 2A, where the fibre is shown as a rod and the cross section in Fig. [Pg 8] 2A has no central canal. The action which takes place is as follows: The cellulose enters into a combination with the alkali and there is formed a sodium cellulose, which has the formula $C_6H_{10}O_5 2NaOH$. This alkali cellulose, however, is not a stable body; by washing with water the alkali is removed, and hydrated cellulose is obtained, which has the formula $C_6H_{10}O_5H_2O$. Water removes the whole of the alkali, but alcohol only removes one half. It has been observed that during the process of washing with water the fibre shrinks very much. This shrinkage is more particularly to be observed in the case of cotton. As John Mercer was the first to point out the action of the alkaline solutions on cotton, the process has become known as "mercerisation".

Solutions of caustic soda of 1.000 or 20° Tw. in strength have very little mercerising action, and it is only by prolonged treatment that mercerisation can be effected. It is interesting to observe that the addition of zinc oxide to the caustic solution increases its mercerising powers. Solutions of 1.225 to 1.275 (that is from 45° to 55° Tw. in strength) effect the mercerisation almost immediately in the cold, and this is the best strength at which to use caustic soda solutions for this purpose. In addition to the change brought about by the shrinking and thickening of the material, the mercerised fibres are stronger than the untreated fibres, and at the same time they have a stronger affinity for dyes, a piece of cloth mercerised taking up three times as much colouring matter as a piece of unmercerised cloth from the same dye-bath.

The shrinkage of the cotton, which takes place during the operation of washing with water, was for a long time a bar to any practical application of the "mercerising" process, but some years ago Lowe ascertained that by conducting the operation while the cotton was stretched or in a state of tension this shrinkage did not take place; further, Thomas and Prevost found that the cotton so treated gained a silky [Pg 9] lustre, and it has since been ascertained that this lustre is most highly developed with the long-stapled Egyptian and Sea Island cottons. This mercerising under tension is now applied on a large scale to produce silkified cotton. When viewed under the microscope, the silkified cotton fibres have the appearance shown in Fig. 3, long rod-like fibres nearly if not quite cylindrical; the cross section of those fibres has the appearance shown in Fig. 3A. This structure fully accounts for the silky lustre possessed by the mercerised fibres. Silky mercerised cotton has very considerable affinity for dye-stuffs, taking them up much more readily from dye-baths, and it is dyed in very brilliant shades.

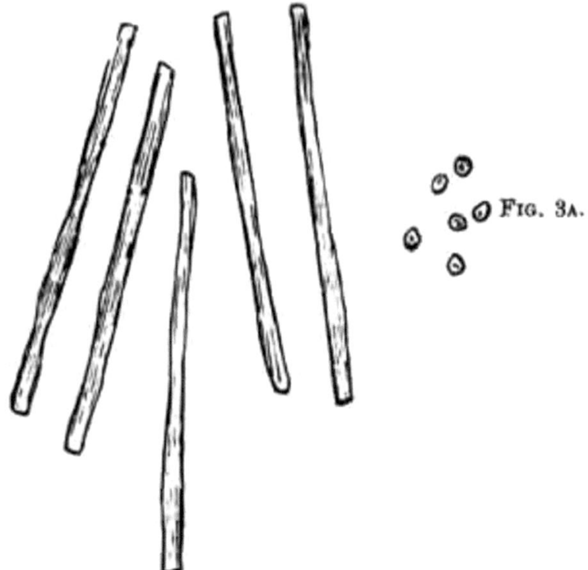

FIG. 3. —
Silkified Cotton Fibre.

In the chapter on Scouring and Bleaching of Cotton, some reference will be made to the action of alkalies on cotton.

ACTION OF ACIDS ON CELLULOSE.

The action of acids on cellulose is a very varied one, being dependent upon several factors, such as the particular acid [Pg 10] used, the strength of the acid, duration of action, temperature, etc. As a rule, organic acids—for example acetic, oxalic, citric, tartaric— have no action on cellulose or cotton. Solutions of sulphuric acid or hydrochloric acid of 2 per cent. strength have practically no action in the cold, and if after immersion the cotton or cellulose be well washed there is no change of any kind. This is important, as in certain operations of bleaching cotton and other vegetable fibres it is necessary to sour them, which could not be done if acids acted on them, but it is important to thoroughly wash the goods afterwards. When the acid solutions are used at the boil they have a disintegrating effect on the cellulose, the latter being converted into hydrocellulose. When dried, the cellulose is very brittle and powdery, which in the case of cotton yarn being so treated would show itself by the yarn becoming tender and rotten. The degree of action varies with the temperature (the higher this is the stronger the action), and also according to the strength of the acid solution. Thus a 10 per cent. solution of sulphuric acid used at a temperature of 80° C. begins to act on cotton after about five minutes' immersion, in half an hour there is a perceptible amount of disintegration, but the complete conversion of the cotton into hydrocellulose requires one hour's immersion. A dilute acid with 8 volumes of water, used in the cold, takes three hours' immersion before any action on the cotton becomes evident.

ACTION OF SULPHURIC ACID ON COTTON.

When cellulose (cotton) is immersed in strong sulphuric acid the cotton becomes gradually dissolved; as the action progresses cellulose sulphates are formed, and some hydrolytic action takes place, with the formation of sugar. This fact has long been known, but only recently has it been shown that dextrose was the variety of sugar which was [Pg 11] formed. On diluting the strong acid solution with water there is precipitated out the hydro or oxycelluloses that have been formed, while the cellulose sulphates are retained in solution.

By suitable means the calcium, barium, or lead salts of these cellulose-sulphuric acids can be prepared. Analysis of them shows that these salts undergo hydrolysis, and lose half their sulphuric acid.

The action of strong sulphuric acid has a practical application in the production of parchment paper; unsized paper is immersed in strong acid of the proper strength for about a minute, and then immediately rinsed in water. The acid acts upon the surface of the paper and forms the cellulose-sulphuric acid which remains attached to the surface. On passing into the water this is decomposed, the acid is washed away, and the cellulose is deposited in an amorphous form on the paper, filling up its pores and rendering it waterproof and grease-proof. Such papers are now largely used for packing purposes.

ACTION OF HYDROCHLORIC ACID.

Dilute hydrochloric acid of from 1° to 2° Tw. in strength, used in the cold, has no action on cellulose. Cotton immersed in acids of the strength named and then well washed in water is not materially affected in any way, which is a feature of some value in connection with the bleaching of cotton, where the material has to be treated at two points in the process with weak acids. Boiling dilute hydrochloric acid of 10° Tw. disintegrates cellulose very rapidly. The product is a white very friable powder, which if viewed under the microscope appears to be fragments of the fibre that has been used to prepare it. The product has the composition $C_{12}H_{22}O_{11}$, and is therefore a hydrate of cellulose, the latter having undergone hydrolysis by taking up the [Pg 12] elements of water according to the equation $2C_6H_{10}O_5 + H_2O = C_{12}H_{22}O_{11}$. By further digestion with the acid, the hydrocellulose, as it is called, undergoes molecular change, and is converted into dextrine. In composition hydrocellulose resembles the product formed by the addition of sulphuric acid which has received the name of amyloid. It differs from cellulose in containing free carboxyl, CO, groups, while its hydroxyl groups, HO, are much more active in their chemical reactions.

Hydrocellulose is soluble in nitric acid, 1.5 specific gravity, without undergoing oxidation. Nitrates are formed varying in composition.

The formation of hydrocellulose has a very important bearing in woollen manufacture. It is practically impossible to obtain wool free from vegetable fibres, and it is often desirable to separate these vegetable fibres. For this purpose the goods are passed into a bath of hydrochloric acid or of weak sulphuric acid. On drying the acid converts the cotton or vegetable fibre into hydrocellulose which, being friable or powdery, can be easily removed, while the wool not having been acted on by the acid remains quite intact. The process is known as "carbonising". It may not only be done by means of the acids named but also by the use of acid salts, such as aluminium chloride, which on being heated are decomposed into free acid and basic oxide. For the same reason it is important to avoid the use of these bodies, aluminium chloride and sulphate, zinc and magnesium chlorides, etc., in the treatment of cotton fabrics; as in finishing processes, where the goods are dried afterwards, there is a great liability to form hydrocellulose with the accompaniment of the tendering of the goods.

ACTION OF NITRIC ACID.

The action of nitric acid on cellulose is a variable one, depending on many factors, strength of acid, duration of [Pg 13] action and temperature. Naturally as nitric acid is a strong oxidising agent the action of nitric acid on cellulose is essentially in all cases that of an oxidant, but the character of the product which is obtained varies very much according to the conditions just noted. When cellulose or cotton in any form is immersed in nitric acid of 1.4 to 1.5 specific gravity for a moment, and the fibre be well washed, there is a formation of hydrate of cellulose which has a gelatinous nature. This is deposited on the rest of the material, which is not materially affected so far as regards strength and appearance, but its power of affinity for dyes is materially increased. There is some shrinkage in the size of the cotton or paper acted upon.

Nitric acid changes all kinds of cellulose into nitro products, the composition of which depends upon the strength of the acid, the duration of treatment, and one or two other factors. The nitrocelluloses are all highly inflammable bodies, the more highly nitrated burning with explosive force. They are produced commercially and are known as "gun cotton" or "pyroxyline". The most highly nitrated body forms the basis of the explosive variety; the least highly nitrated forms that of the soluble gun cotton used for making collodion for photographic and other purposes.

The products formed by the action of nitric acid are usually considered to be nitrocelluloses. It would appear that they are more correctly described as cellulose-nitrates, for analysis indicates the presence of the NO_3 group, which is characteristic of nitrates, and not of the NO_2 group, which is the feature of nitro bodies in general. Further, nitro compounds, when subject to the action of reducing agents, are converted into amido compounds, as is the case, for instance, with nitro-benzene, $C_6H_5NO_2$, into aniline, $C_6H_5NH_2$, or with nitro-naphthalene, $C_{10}H_7NO_2$, which changes into naphthylamine, [Pg 14] $C_{10}H_7NH_2$.

But the nitric acid derivatives of cellulose are not capable of conversion by reducing agents into similar amido compounds. They have the following properties, which accord more closely with nitrates than with nitric bodies: alkalies remove the nitric acid; cold sulphuric acid expels the nitric acid, cellulose sulphates being formed; boiling with ferrous sulphate and hydrochloric acid causes the elimination of the nitric acid as nitric oxide (on which reaction a method for determining the degree of nitration of gun cotton is based). It is best therefore to consider them as cellulose nitrates. Several well-characterised cellulose nitrates have been prepared, but is an exceedingly difficult matter to obtain any one in a state of purity, the commercial articles being always mixtures of two or three. Those that are best known and of the most importance are the following: —

Cellulose Hexa-nitrate, $C_6H_4O_5(NO_3)_6$. This forms the principal portion of the commercial explosive gun cotton, and is made when a mixture of strong nitric acid and strong sulphuric acid is allowed to act on cotton at from 50 to 55° F. for twenty-four hours. The longer the action is prolonged, the more completely is the cotton converted into the nitrate, with a short duration the finished product contains lower nitrates. This hexa-nitrate is insoluble in ether, alcohol, or in a mixture of those solvents, likewise in glacial acetic acid or in methyl alcohol.

Cellulose Penta-nitrate, $C_6H_5O_5(NO_3)_5$, is found in explosive gun cotton to a small extent. When gun cotton is dissolved in nitric acid and sulphuric acid is added, the penta-nitrate is thrown down as a precipitate. It is not soluble in alcohol, but is so in a mixture of ether and alcohol, it is also slightly soluble in acetic acid. Solutions of caustic potash convert it into the di-nitrate.

Cellulose Tetra-nitrate, $C_6H_6O_5(NO_3)_4$, and Cellulose Tri-nitrate, $C_6H_7O_5,(NO_3)_3$, form the basis of the pyroxyline or solu [Pg 15] ble gun cotton of commerce. It has not been found possible to separate them owing to their behaviour to solvents being very similar. These nitrates are obtained by treating cotton with nitric acid for twenty or thirty minutes. They are characterised by being more soluble than the higher nitrates and less inflammable. They are freely soluble in a mixture of ether and alcohol, from which solutions they are precipitated in a gelatinous form on adding chloroform. Acetic ether, methyl alcohol, acetone and glacial acetic acid, will also dissolve these nitrates.

Cellulose Di-nitrate, $C_6H_8O_5(NO_3)_2$, is obtained when cellulose is treated with hot dilute nitric acid, or when the high nitrates are boiled with solutions of caustic soda or caustic potash. Like the last-mentioned nitrates it is soluble in a mixture of alcohol and ether, in acetic ether, and in absolute alcohol. The solution of the pyroxyline nitrates in ether and alcohol is known as collodion, and is used in photography and in medical and surgical work.

One of the most interesting applications of the cellulose nitrates is in the production of artificial silk. Several processes, the differences between which are partly chemical and partly mechanical, have been patented for the production of artificial silk, those of Lehner and of Chardonnet being of most importance. They all depend upon the fact that when a solution of cellulose nitrate is forced through a fine aperture or tube, the solvent evaporates almost immediately, leaving a gelatinous thread of the cellulose nitrate which is very tough and elastic, and possesses a brilliant lustre. Chardonnet dissolves the cellulose nitrate in a mixture of alcohol and ether, and the solution is forced through fine capillary tubes into hot water, when the solvents immediately evaporate, leaving the cellulose nitrate in the form of very fine fibre, which by suitable machinery is drawn away as fast as it is formed. Lehner's process [Pg 16] is very similar to that of Chardonnet. Lehner uses a solution of cellulose nitrate in ether and alcohol, and adds a small quantity of sulphuric acid; by the adoption of the latter ingredient he is able to use a stronger solution of cellulose nitrate, 10 to 15 per cent., than would otherwise be possible, and thereby obtains a stronger thread which resists the process of drawing much better than is the case when only a weak solution in alcohol and ether is employed. By subsequent treatment the fibre can be denitrated and so rendered less inflammable.

The denitrated fibres thus prepared very closely resemble silk in their lustre; they are not quite so soft and supple, nor are they in any way so strong as ordinary silk fibre of the same diameter.

Artificial silk can be dyed in the same manner as ordinary silk.

ACTION OF OXIDISING AGENTS ON CELLULOSE OR COTTON

Cellulose resists fairly well the action of weak oxidising agents; still too prolonged an action of weak oxidising agents has some influence upon the cotton fibre, and it may be worth while to point out the action of some bodies having an oxidising effect.

Nitric acid of about 1.15 specific gravity has little action in the cold, and only slowly on it when heated. The action is one of oxidation, the cellulose being transformed into a substance known as oxycellulose. This oxycellulose is white and flocculent. It tends to form gelatinous hydrates with water, and has a composition corresponding to the formula $C_6H_{10}O_6$. It is soluble in a mixture of nitric and sulphuric acids, and on diluting this solution with water a trinitrate precipitates out. A weak solution of soda dissolves this oxycellulose with a yellow colour, while strong sulphuric acid forms a pink colouration. It is important to note that [Pg 17] nitric acid of the strength given does not convert all the cellulose into oxycellulose, but there are formed also carbonic and oxalic acids. When cotton is passed into strong solutions of bleaching powder and of alkaline hypochlorites and then dried, it is found to be tendered very considerably. This effect of bleaching powder was first observed some thirteen years ago by George Witz, who ascribed the tendering of the cotton to the formation of an oxycellulose. Although the composition of this particular oxycellulose so formed has not yet been ascertained, there is reason to think that it differs somewhat from the oxycellulose formed by the action of the weak nitric acid. A notable property of the oxycellulose now under consideration is its affinity for the basic coal-tar dyes, which it will absorb directly. The oxycellulose is soluble in alkaline solutions.

In the ordinary bleaching process there is considerable risk of the formation of oxycellulose by the employment of the bleaching solutions of too great a strength, or in allowing the goods to lie too long before the final washing off. The presence of any oxycellulose in bleached cotton may be readily determined by immersing it in a weak solution of Methylene blue, when, if there be any oxycellulose present, the fibre will take up some of the dye-stuff.

Permanganate of potash is a very powerful oxidising agent. On cellulose neutral solutions have but little action, either in the cold or when heated. They may, therefore, be used for the bleaching of cotton or other cellulose fibres.

Alkaline solutions of permanganate convert the cellulose into oxycellulose, which resembles the oxycellulose obtained by the action of the nitric acid.

Chromic acid, when used in the form of a solution, has but little action on cellulose. In the presence of mineral acids, and used warm or boiling, chromic acid oxidises cellulose into oxycellulose and other products. [Pg 18]

It is therefore always advisable in carrying out any technical process connected with cotton which involves its treatment with oxidising agents of any kind, and where it is desired not to alter the cotton, to allow these actions to be as short as possible.

Dyes and Cotton Dyeing.—An account of the chemistry of the cotton fibre would not be complete unless something is said about the reactions involved in the processes of dyeing and printing cotton. This is a most interesting subject and opens up quite a number of problems relating to the combination of the fibre with colouring matter of various kinds, but here only a brief outline of the principles that present themselves in considering the behaviour of the cotton fibre as regards colouring matter will be given.

When the question is considered from a broad point of view, and having regard to the various affinities of the dyes for cotton; we notice (1) that there is a large number of dye-stuffs—the Benzo, Congo, Diamine, Titan, Mikado, etc., dyes—that will dye the cotton from a plain bath or from a bath containing salt, sodium sulphate, borax or similar salts; (2) that there are dyes which, like Magenta, Safranine, Auramine and Methyl violet, will not dye the cotton fibre direct, but require it to be mordanted or prepared with tannic acid; (3) that there are some dyes or rather colouring matters which, like Alizarine, Nitroso-resorcine, barwood, logwood, etc., require alumina, chrome and iron mordants; (4) that there are some dyes which, like the azo scarlet and azo colours in general, cannot be used in cotton dyeing; (5) that there are a few dyes, *i.e.*, indigo, which do not come under this grouping.

From the results of recent investigations into the chemistry of dyeing it is now considered that for perfect dyeing to take place there must be formed on the fibre a combination which is called a "colour lake," which consists of at least two constituents; one of these is the dye-stuff or the colour [Pg 19] ing matter itself, the other being either the fibre or a mordant, if such has to be used. The question of the formation of colour lakes is one connected with the molecular constitution of the colouring matter, but much yet remains to be done before the proper functions and mode of action of the various groups or radicles in the dye-stuffs can be definitely stated. While the constitution of the dye-stuff is of importance, that of the substance being dyed is also a factor in the question of the conditions under which it is applied.

In dealing with the first of the above groups of dyes, the direct dyes, the colourist is somewhat at a loss to explain in what manner the combination with the cotton fibre is brought about. The affinity of cellulose for dyes appears to be so small and its chemical activities so weak, that to assume the existence of a reaction between the dye-stuff and the fibre, tending to the formation of a colour lake, seems to be untenable. Then, again, the chemical composition and constitution of the dyes of this group are so varied that an explanation which would hold good for one might not do so for another. The relative fastness of the dyes against washing and soaping precludes the idea of a merely mechanical absorption of the dye by the fibre; on the other hand the great difference in the fastness to soaping and light between the same dyes on cotton and wool would show that there has not been a true formation of colour lake.

The dyeing of cotton with the second group of dyes is more easily explained. The cotton fibre has some affinity for the tannic acid used in preparing it and absorbs it from the mordanting bath. The tannic acid has the property of combining with the basic constituents of these dyes and forms a true colour lake, which is firmly fixed on the fibre. The colour lake can be formed independently of the fibre by bringing the tannic acid and the dye into contact with one another. [Pg 20]

In the case of the dyes of the third group, the formation of a colour lake between the metallic oxide and the colouring matter can be readily demonstrated. In dyeing with these colours the cotton is first of all impregnated with the mordanting oxide, and afterwards placed in the dye-bath, the mordant already fixed on the fibre then reacts with the dye, and absorbs it, thus dyeing the cotton. To some extent the dyeing of cotton with the basic dyes of the second group and the mordant dyes of the third group is almost a mechanical one, the cotton fibre taking no part in it from a chemical point of view, but simply playing the part of a base or foundation on which the colour lake may be formed. In the case of the dyes of the fourth group, there being no chemical affinity of the cotton known for them, these dyes cannot be used in a successful manner; cotton will, if immersed in a bath containing them, more or less mechanically take up some of the colour from the liquor, but such colour can be almost completely washed out again, hence these dyes are not used in cotton dyeing, although many attempts have been made to render them available.

Indigo is a dye-stuff that stands by itself. Its combination with the cotton fibre is chiefly of a physical rather than a chemical nature; it does not form colour lakes in the same way as Alizarine and Magenta do.

Cellulose can be dissolved by certain metallic solutions and preparations:—

(1) **Zinc Chloride.**—When cotton or other form of cellulose is heated with a strong solution, 40 to 50 per cent., it slowly dissolves to a syrupy liquid. On diluting this liquid with water the cellulose is thrown down in a gelatinous form, but more or less hydrated, and containing some zinc oxide, 18 to 25 per cent., in combination.

(2) **Zinc Chloride and Hydrochloric Acid.**—When zinc chloride is dissolved in hydrochloric acid a liquid is ob [Pg 21] tained which dissolves cellulose; on dilution the cellulose is re-precipitated in a hydrated form. It is worth noting that the solution is not a stable one: on keeping, the cellulose changes its character and undergoes hydrolysis to a greater or less extent.

(3) **Ammoniacal Copper.**—When ammonia is added to a solution of copper sulphate, there is formed at first a pale blue precipitate of copper hydroxide, which on adding excess of ammonia dissolves to a deep blue solution—a reaction highly characteristic of copper. The

ammoniacal copper solution thus prepared has, as was first observed by John Mercer, the property of dissolving cellulose fairly rapidly, even in the cold.

If instead of preparing the ammoniacal copper solution in the manner indicated above, which results in its containing a neutral ammonium salt, the copper hydroxide be prepared separately and then dissolved in ammonia a solution is obtained which is stronger in its action.

The cupra-ammonium solutions of cellulose are by no means stable but change on keeping. When freshly prepared, the cellulose may be precipitated from them almost unchanged by the addition of such bodies as alcohol, sugar and solutions of neutral alkaline salts. On keeping the cellulose undergoes more or less hydrolysis or even oxidation, for it has been observed that oxycellulose is formed on prolonged digestion of cellulose with cupra-ammonium solutions, while there is formed a fairly large proportion of a nitrite.

On adding lead acetate to the cupra-ammonium solution of cellulose, a compound of lead oxide and cellulose of somewhat variable composition is precipitated. It is of interest also to note that on adding metallic zinc to the cupra-ammonium solution the copper is thrown out and a solution containing zinc is obtained.

This action of cupra-ammonium solutions on cellulose has [Pg 22] been made the basis for the production of the "Willesden" waterproof cloths. Cotton cloths or paper are passed through these solutions of various degrees of strength according to requirements, they are then passed through rollers which causes the surface to become more compact. There is formed on the surface of the goods a deposit of a gelatinous nature which makes the surface more compact, and the fabric becomes waterproof in character while the copper imparts to them a green colour and acts as a preservative. The "Willesden" fabrics have been found very useful for a variety of purposes. [Pg 23]

CHAPTER II.

SCOURING AND BLEACHING OF COTTON.

Preparatory to the actual dyeing operations, it is necessary to treat cotton in any condition—loose cotton, yarn, or piece—so that the dyeing shall be properly done. Raw cotton contains many impurities, mechanical and otherwise; cotton yarns accumulate dirt and impurities of various kinds during the various spinning operations, while in weaving a piece of cotton cloth it is practically impossible to keep it clean and free from dirt, etc. Before the cotton can be dyed a perfectly level and uniform shade, free from dark spots or light patches, these impurities must be removed, and therefore the cotton is subjected to various scouring or cleansing operations with the object of effecting this end. Then again cotton naturally, especially Egyptian cotton, contains a small quantity of a brown colouring matter, and this would interfere with the purity of any pale tints of blue, rose, yellow, green, etc., which may be dyed on the cotton, and so it becomes necessary to remove this colour and render the cotton quite bright. This is commonly called "bleaching". It is these preparatory processes that will be dealt with in this chapter.

Scouring Cotton.—When dark shades—blacks, browns, olives, sages, greens, etc., are to be dyed it is not needful to subject the cotton to a bleaching operation, but simply to a scouring by boiling it with soda or caustic soda. This is very often-carried out in the same machine as the goods will [Pg 24] be dyed in; thus, for instance, in the case of pieces, they would be charged in a jigger, this would be filled with a liquor made from soda or from caustic soda, and the pieces run from end to end, while the liquor is being heated to the boil—usually half to three-quarters of an hour is sufficient. Then the alkali liquor is run out, clean water run into the jigger and the pieces washed, after which the dyes, etc., are run into the jigger and the dyeing done. There is usually used 2 lb. to 3 lb. of caustic soda, or 3 lb. to 4 lb. of soda for each 100 lb. of goods so treated.

If the ordinary dyeing machines are not used for this purpose, then the ordinary bleachers' kiers may be used. These will be described presently.

Bleaching of Cotton.—Cotton is bleached in the form of yarn, or in the finished pieces. In the latter case the method depends very largely on the nature of the fabric; it is obvious that fine fabrics, like muslins or lace curtains, cannot stand the same rough treatment as a piece of twilled calico will. Then, again, the bleaching process is varied according to what is going to be done with the goods after they are bleached; sometimes they are sent out as they leave the bleach-house; again, they may have to be dyed or printed. In the first case the bleach need not be of such a perfect character as in the last case, which again must be more perfect than the second class of bleach. There may be recognised:—

(1) Market or white bleach. (2) Dyers or printers' bleach. (3) Madder bleach.

As the madder bleach is by far the most perfect of the three, and practically includes the others, this will be described in detail, and differences between it and the others will be then pointed out. A piece is subjected to the madder [Pg 25] bleach which has afterwards to be printed with madder or alizarine. Usually in this kind of work the cloths are printed with mordant colours, and then dyed in a bath of the dye-stuff. This stains the whole of the piece, and to rid the cloth of the stain where it has to be left white, it is subjected to a soap bath. Now, unless the bleach has been thorough, the whites will be more or less stained permanently, and to avoid this cloths which are to be printed with alizarine colours are most thoroughly bleached. The madder bleach of the present day generally includes the following series of operations:—

(1) Stitching. (2) Singeing. (3) Singeing wash. (4) Lime boil. (5) Lime sour. (6) Lye boil. (7) Resin boil. (8) Wash. (9) Chemicing. (10) White sour.

(1) **Stitching.**—The pieces are fastened together by stitching into one long rope, which is passed in a continuous manner through all operations in which such a proceeding is possible. This stitching is done by machines, the simplest of which is the donkey machine, whereby the ends of the pieces, which are to be stitched together, are forced by a pair of cogwheels working together on to the needle carrying a piece of thread, this is then pulled through and forms a running stitch, a considerable length of thread being left on each side so as to prevent as far as possible the pulling asunder of the pieces by an accidental drawing out of the thread.

Birch's sewing machine is very largely used in bleach works. It consists essentially of a Wilcox & Gibb machine fitted on a stand so as to be driven by power. The pieces are carried under the needle by a large wheel, the periphery of which contains a number of projecting pins that, engaging in the cloth, carry it along. [Pg 26]

There is also a contrivance by which these pieces to be sewn can be kept stretched, this takes the form of an arm with clips at the end, which hold one end of the cloth while it is running through the machine. The clip arrangement is automatic, and just before the end passes under the needle it is released, and the arm flies back ready for the next piece; it is, however, not necessary to use this arm always. This machine gives a chain stitch sufficiently firm to resist a pull in the direction of the length of the pieces, but giving readily to a pull at the end of the thread.

The Rayer & Lincoln machine is an American invention, and is much more complicated than Birch's. It consists of a sewing machine mounted on the periphery of a large revolving wheel. This carries a number of pins, which, engaging in the cloth to be stitched, carry it under the needle of the machine. Besides sewing the pieces together this machine is fitted with a pair of revolving cutters which trim the ends of the pieces as they pass through in a neat clean manner. There is also an arrangement to mark the pieces as they are being stitched. Like Birch's it produces a chain stitch.

What is important in sewing the ends of pieces together is to get a firm uniform stitch that lies level with the cloths without any knots projecting, which would catch in the bleaching machinery during the processes of bleaching, and this might lead to much damage being done.

Should it be necessary to mark the pieces so that they can be recognised after bleaching, the best thing to use is printers' ink. Gas tar is also much used, and is very good for the purpose. Coloured inks do not resist the bleaching sufficiently well to be used satisfactory. Vermilion and Indian red are used for reds, yellow ochre is the fastest of the yellows, there is no blue which will stand the process, and Guignet's green is the only green that will at all resist [Pg 27] the process, umber will serve for brown. All these colours are used in the form of printing ink.

The next operation is a very important one, which cannot be too carefully carried out, that is: —

(2) **Singeing.**—For printing bleaches the cloths are singed. This has for its object the removal from the surface of the cloth of the fine fibres with which it is covered, and which would, if allowed to remain, prevent the designs printed on from coming out with sufficient clearness, giving them a blurred appearance.

Singeing is done in various ways, by passing the cloth over a red-hot copper plate, or over a red-hot revolving copper cylinder, or through a coke flame, or through gas flames, and more recently over a rod of platinum made red hot by electricity.

Plate singeing is the oldest of these methods and is still largely used. In this method a semi-cylindrical copper plate is heated in a suitable furnace to a bright red heat, the cloths are rapidly passed over it, and the loose fibres thereby burnt off. One great trouble is to keep the plate at one uniform heat over the whole of its surface, some parts will get hotter than others, and it is only by careful attention to the firing of the furnace that this can be obtained. To get over these difficulties Worral introduced a roller singeing machine in which the plate was replaced by a revolving copper roller, heated by a suitable furnace; the roller can be kept at a more uniform temperature than the plate. The singe obtained by the plate and roller is good, the principal fault being that if the cloths happen to get pressed down too much on the hot plate the loose ends are not burnt off as they should be. With both plate and roller the cloths are singed only on one side, and if both sides require to be singed a second passage is necessary. Both systems still retain their hold as the principal methods in use, notwithstanding the introduction of more modern methods. [Pg 28]

Singeing by passing the cloths over a row of Bunsen burners has come largely into use. This has the great advantage of being very cleanly, and of doing the work very effectually, much more thoroughly than any other method, which is due to the fact that while in the methods described above only the loose fibres on the surface are burnt off; with gas all the loose fibres are burnt off. This is brought about by the gas flame passing straight through the cloth. It is not necessary to describe the gas singeing machine in detail. Singeing machines should be kept scrupulously clean and free from fluff, which is liable to collect round them, and very liable to fire. Some machines are fitted with a flue having a powerful draught which carries off this fluff, away from any source of danger.

(3) **Singeing Wash.**—After being singed the cloths are run through a washing machine to remove by water as much of the loose charred fibres as possible. The construction of a washing machine is well known. It consists of a pair of large wooden rollers set above a trough containing water and into which a constant stream of water flows. In the trough is also fixed another wooden roller and the pieces are passed round this bottom roller and between the top rollers. The cloth is passed through and round the rollers several times in a spiral form so that it passes through the water in the trough frequently, which is a great advantage, as the wash is thus much more effectual. The pressure between the two top rollers presses out any surplus water. The operation scarcely needs any further description.

(4) **Lime Boil.**—After the cloth leaves the singeing or grey wash, as it is often called, it passes through the liming machine, which is made very similar to the washing machine. In this it passes through milk of lime, which should be made from freshly slaked lime. The latter maybe prepared in a pasty form in a stone cistern. The lime used should be of [Pg 29] good quality, free from stones, badly burnt pieces or any other insoluble material, so that when slaked it should give a fine smooth pasty mass.

Lime should not be slaked too long before using, as it absorbs carbonic acid from the atmosphere, whereby carbonate of lime is formed, and this is useless for liming cloth. The pasty slaked lime may be mixed with water to form the milk of lime, and this can be run from the cistern in which it is prepared into the liming machine as it is required; the supply pipe should be run into the bottom of the trough of the liming machine and not over the top, in which latter case it may splash on to the cloths and lead to overliming, which is not to be desired on account of its liability to rot the cloth. The amount of lime used varies in different bleachworks, and there is no rule on the subject; about 5 lb. to 7 lb. of dry lime to 100 lb. of cloth may be taken as a fair quantity to use.

The lime boil has for its object the removal or rather the saponification of the resinous and fatty matters present in the grey cloth, either naturally or which have been added in the process of weaving, or have got upon the cloth accidentally during the processes of spinning and weaving. With these bodies the lime forms insoluble lime soaps; these remain in the cloth, but in a form easily decomposable and removable by treatment with acids and washing. Soda or potash is not nearly so good for this first boiling as lime—for what reason is somewhat uncertain, but probably because they form with the grease in the cloths soluble soaps, which might float about the kier and accumulate in places where they are not required and thus lead to stains, whereas the insoluble lime soap remains where it was formed. The lime also seems to attack the natural colouring matter of the cotton, and although the colour of the limed cloth is darker than before boiling, yet the nature of the colour is so altered that it is more easily removed in the after processes. Besides [Pg 30] these changes the starchy matters put into the cloth in the sizing are dissolved away. Great care should be taken to see that the goods are evenly laid in the kiers, not too tight, or the liquor will not penetrate properly; and not too slack, or they will float about and get entangled and more or less damaged. Then again care should be taken, especially when using low-pressure kiers, to see that the supply of liquor does not get too low, in which case the goods in the upper part of the kier are liable to get dry and are tendered thereby. So long as the goods in the kiers are not allowed to get dry there is no risk of damage; this trouble rarely arises with the Barlow and injector kiers. The inside of the kiers should be kept well limed, so that the goods shall not come in contact with the bare iron or metal of which the kier is constructed, as this would be very likely to lead to stains being produced which are by no means easy to remove. It is usual, and it is a good plan with almost all kinds of kiers, except the Mather and Edmeston kiers, to put a number of large pebbles or small stones at the bottom of the kier, which serves to make a false bottom on which the goods rest and through which the liquor penetrates and flows away. Before using, the stones should be well washed to free them from dirt and grit.

The lime boil is carried out in what are called "kiers". Many forms of kiers have been devised, but the one in most general use is that known as the "injector kier," of which a drawing is given in Fig. 4, of the form made by Messrs. Mather & Platt of Salford. Injector kiers are made to work either under a pressure of 40 lb. to 50 lb. of steam per square inch, when they are called high-pressure kiers, or at a pressure of 15 lb. to 20 lb., when they are called low-pressure kiers. The one shown in the drawing is intended for low-pressure kiers. The principle of construction is the same in all, the details varying somewhat with different makers. Injector [Pg 31]

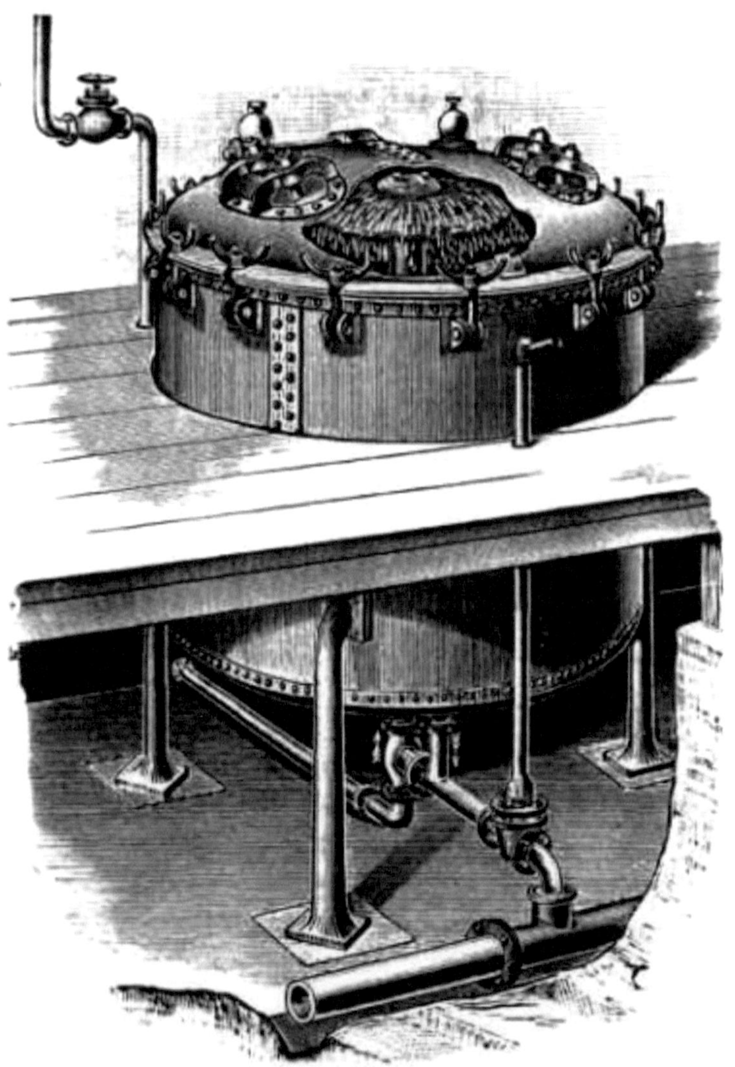

Fig. 4.—Mather & Platt's Low-pressure Bleaching Kier.

kiers consist of a hollow, upright iron cylinder made of plates riv-
eted together; the top is made to lift off, but can be fastened down
tightly by means of bolts and nuts as shown in the drawing. From
the bottom, and placed centrally, rises a pipe, known as the puffer

pipe; this terminates at the [Pg 32] top in a rose arrangement. The lower end of the pipe is perforated. A jet of steam is sent in at the bottom of this pipe, and by its force any liquor at the bottom of the kier is forced up the puffer pipe and distributed in a spray over any goods which may be in the kier. The liquor ultimately finds its way to the bottom of the kier ready to be blown up again. This circulation of the liquor can be maintained for any length of time and through its agency every part of the goods gets thorough and effectual treatment.

The length of the boil depends upon the kiers; with the open kier about ten hours are usually given, with the Barlow and injector kiers, working at a pressure of 40 lb. to 50 lb., six to seven hours are given.

(5) **Lime or Grey Sour.**—After the lime boil, the next operation is that of the lime sour or grey sour as it is often called to distinguish it from a subsequent souring. The souring is done in a machine constructed in the same way as a washing machine; the trough of the machine is filled with hydrochloric acid at 2° Tw., which is kept ready prepared in a stone cistern and run into the machine as required (it is not advisable to use acid stronger than this). After passing through the sour the goods are piled in a heap on the stillage for a few hours. The acid attacks the lime soap which was formed during the lime boil, decomposes it and dissolves out the lime with the formation of calcium chloride, while the fat of the soap is liberated, the former is washed away in the subsequent washing, while the latter remains to a large extent on the goods, and is removed by the lye boil that follows. Sulphuric acid is not so satisfactory to use for the lime sour as hydrochloric acid, because it forms with the lime the insoluble sulphate of calcium, which is difficult to entirely remove from the goods, whereas the chloride is very soluble and is entirely eliminated from the goods by the washing that follows. [Pg 33]

It is advisable to keep the acid at a uniform strength in the machine. The Twaddell is here of no use as an indicator of the actual strength, because the lime which the acid dissolves, while it neutralises and reduces the strength of the acid, actually raises the Twaddell, under which circumstance the only safe method is a chemical test. This can be carried out very simply and with a sufficient degree of accuracy by the workmen, and if it be done at regular intervals during the souring, and the supply of the fresh acid be regulated, the sour will be kept at a more uniform strength and more uniform results will be obtained than if the souring were done in a more empirical fashion. The test is best and most easily done as follows:—

Prepare a solution of 1 oz. of the powdered high strength 98 per cent. caustic soda in 1 pint of water, weighing and measuring these quantities very carefully. Now take a tall, narrow, white bottle of about 5 oz. capacity and make a mark on the neck. Fill this bottle with the test solution.

Now take exactly 5 ozs. of freshly prepared sour of 2° Tw., pour into a jar, and add carefully some of the soda-test solution until a piece of cloth dyed with turmeric is turned brown, when the acid is neutralised. Now make a mark on the bottle of soda to show how much has been used. In all subsequent tests of the sour 5 ozs. should always take the same quantity of soda solution; if it takes less it is too weak, if more it is too strong; the remedy in each case is obvious. It is worth while to graduate the test bottle for 1°, 3°, 4°, 5° Twaddell, as well as for 2° Tw. acid.

After the souring it is often the custom to pile the goods on to a wooden stillage, but the goods should not be left too long so piled up for they may become dry, either entirely or in parts. In any case, as the goods dry the acid becomes concentrated and attacks them and makes them tender, which is not at all desirable. Therefore, if it is not convenient [Pg 34] to proceed with them for some time after souring, they should be moistened with water from time to time, but it is best to wash them off at once, whereby they are made ready for the next operation.

(6) **Lye Boil**.—This is, perhaps, the most important operation in the whole process of bleaching, especially if the cloths are going to be printed in the so-called madder style with alizarine colours, or otherwise stains are liable to occur in the final stage, and it is then sometimes difficult to put the blame for these upon the right shoulders.

In principle the lye boil is simple, consisting in boiling the goods with a solution of soda ash, or caustic soda. The quantity of ash used varies in different works, as might naturally be expected; from 170 lb. to 200 lb. of ash to 10,000 lb. of cloth is a fair proportion to use. The length of boil averages about four hours, certainly not less than three should be given, and it is not necessary to give more than five hours in either ordinary kiers, with central puffer pipe, or in injector kiers.

Care should be taken to see that the goods are well packed into the kiers, not too tightly or the lye will fail to penetrate equally all through, and this is important if a uniform bleach is desired; neither should they be too loose, or they will float about and get torn. It is not necessary to be particular about the quantity of water used, except that it must be sufficient to keep the goods well covered, and still have enough to keep the circulation energetic. When the water is not sufficient in amount the goods get somewhat dry; there is then a liability to tendering, but with plenty of water there is no fear of any damage being done during a boil with alkali. Some works use caustic soda instead of soda ash in which case less is required, from 120 lb. to 150 lb. to 10,000 lb. of cloth, otherwise no alteration is made in the mode of boiling. [Pg 35]

This lye boil clears away the fatty and waxy matter left in the goods after the lime sour, and thus prepares the way for the next boil. There is no advantage in using caustic soda in this preliminary boil, soda ash being just as effective and cheaper.

(7) **Resin Boil.**—Following the lye boil is the resin boil which consists in boiling the goods in a resin soap liquor. This is made as follows: a soda ash liquor of about 15° to 20° Tw. is prepared, and into this is thrown resin, broken up into small pieces.

The whole is boiled up until the resin is dissolved, and then as much more is added in small quantities as the alkali will take up. The soda liquor should not be much weaker than 20° Tw., it will then be heavier than the resin which will float on the top, it is found to dissolve quicker and better than when the liquor is weak, in which case, the resin would sink to the bottom of the boiler and would there melt into a single mass difficult to dissolve. The resin soap liquor when made is ready to be used. The proportions of resin and alkali used in the boil vary in different works, but, as a rule, the quantities for 10,000 lb. of goods are 430 lb. of 58 per cent. soda ash, 180 lb. of resin, and 80 lb. of 70 per cent. caustic soda. Too much resin should be avoided, as it is found that with an excess the whites obtained are not nearly so good as when the right quantity is used; on the other hand, too little acts much in the same way. It may be taken that from 1½ to 1¾ per cent. of the weight of the goods is about the right proportion; 1 per cent. being too little, and 2 per cent. too much. The quantity of soda used should be rather more than twice that of the resin, from 3½ to 4 per cent. The length of boil is usually about twelve hours in a low-pressure kier; in a high-pressure kier about seven hours is sufficient.

What the special function of the resin is in this boil is [Pg 36] not definitely known; but experience, both on a large and small scale, proves that it is essential to obtaining a good white for alizarine printing; without it, when the goods are dyed with alizarine after the mordants have been printed on, they frequently take a brown stain — with the resin this never or but rarely happens.

(8) **Wash.** — After the lye boils the goods must be washed, and it is important that this be done in as thorough a manner as possible. With the object of accomplishing this most thoroughly many washing machines have been invented, the main idea in all being to bring every part of the goods into contact with as much water as possible. Bridson's is an old form, and a very good one, the principle of this machine is to cause the cloth to pass to and fro, and to flap upon the surface of the water in the trough of the machine.

Furnival's square beater works on much the same principle, and does its work effectively. More modern washing machines are those of Birch, Farmer, Mather & Platt, and Hawthorne, where by the peculiar construction of the rollers and the use of beaters the cloth is very effectually washed. These machines are much more economical in the use of water than the older forms, and yet they do their work as well, if not better.

(9) **Chemicing.** — This is the actual bleaching operation, familiarly known as "chemicing," that is, the treatment of the goods with bleaching powder. The previous operations have resulted in obtaining a cloth free from grease, natural or acquired, and from other

impurities, but it still has a slight brownish colour. This has to be removed before the goods can be considered a good white, which it is the aim of every bleacher they should be.

To get rid of this colour they are subjected to some final operations, the first of which is now to be considered. The chemicing consists in running the goods through a weak [Pg 37] solution of bleaching powder (chloride of lime), piling the goods up into heaps, and allowing them to lie overnight, the next day they are finished. As the cloth has received, or ought to have received, a thorough bottoming, only a weak bath of chemic is required, about ½ to 1° Tw. is quite sufficient. The solution is prepared in a stone cistern. There is very little difficulty in making it, the only precaution necessary is to have the solution quite clear and free from undissolved particles, for if these get upon the cloth they will either lead to the production of minute holes, or they may overbleach the fibre, which in such case will have the power of attracting excess of colour in any subsequent dyeing process and thus lead to stains, the origin of which may not be readily grasped at the first sight.

It is best, therefore, either to allow the solution to settle in the cistern till quite clear, which is the simplest way, or to filter through cloth.

The chemicing is best done cold and with weak solution, at ½° Tw. rather than 1° Tw. Warming the liquor increases the rapidity of the bleaching action, but there is a risk of over-chloring, which must be avoided as far as possible, because there is then danger of tendering the fibre, moreover, such over-chlored cloth has an affinity for colouring matters that is not at all desirable, as it leads to the production of stains and patches in the dyeing operations. It is much better, when a single chemicing does not bleach the cloth sufficiently and give a white, to run the cloth twice through a weak liquor rather than once through a strong liquor.

Although the chemicing is followed by a sour, which, acting on the bleaching powder, liberates chlorine that bleaches the fabric, yet the greatest proportion of the bleaching effect is brought about while the pieces are being piled up into heaps between the chemicing and the sour. In this state they should be left for some hours, covered [Pg 38] over with a damp sheet, care being taken that they are not left piled so long as to become dry, as in this event there is a great risk of tendering the cloth or fabric; it is, therefore, a good plan to moisten them with a little water from time to time. They should not be tightly piled up, but be as loose as possible, so that the air can get to them, as it is the carbonic acid and other acid vapours in the air, that by acting on the chemic causes slow liberation of chlorine, which effects the bleaching of the goods.

(10) **White Sour**.—After the chemicing the goods are treated to a sour, for which purpose either hydrochloric acid or sulphuric acid may be used.

Hydrochloric acid possesses the advantage of forming a more soluble salt of lime (calcium chloride) than does sulphuric acid (calcium sulphate), and it has a more solvent action upon any traces of iron and other metallic oxide stains which may be present in the goods. On the other hand, on account of its fuming properties, it is unpleasant to work with. The souring is done by passing the goods through an acid liquor at 2° Tw. strong and piling for two or three hours, after which it is washed. This final washing must be thorough, so that all traces of acid and chemic are washed out, otherwise there is a tendency for the goods to acquire a yellowish colouration.

So far the routine has been described of the so-called madder bleach, the most perfect kind of bleach applied to cotton cloths. Besides this two other kinds of bleach are distinguished in the trade. Turkey red and market bleach. The former is used when the cloth or yarn is to be dyed plain or self-coloured with delicate shades with Alizarine; the latter is used for cloth sold in the white. As the operations involved in producing these are identical in their method of manipulation to those already described, it will only be necessary to give an outline of the process for each one. [Pg 39]

Turkey Red Bleach—(1) Rinse through water into a kier and boil for two hours. (2) Lime boil for three to four hours. The amount of lime required is rather less than what is used with the madder bleach, from 2½ lb. to 3 lb, lime to each 1 cwt. of goods being quite sufficient. (3) Souring as in the madder bleach. (4) Lye boil, using about 100 lb. caustic soda to a ton of goods, and giving ten hours' boil. (5) Second lye boil using about 50 lb. soda ash to a ton of goods, after which the goods are well washed. (6) Chemicing as with the madder bleach. (7) Souring as with the madder bleach, then washing well.

This represents an average process, but almost every bleacher has his own methods, differing from the above in some of the details and this applies to all bleaching processes. It is obvious that the details may be varied to a great extent without changing the principles on which the process depends.

Market Bleach—Here all that requires to be done is to get the cloth of a sufficient degree of whiteness to please the eye of the customer. Market bleachers have, however, to deal with a wider range of goods than is dealt with in the former kinds of bleaches, from very fine muslins to very heavy sheetings. Now it is obvious from a merely mechanical point of view, that the former could not stand as rough a process as the latter, therefore there must be some differences in the details of muslin bleaching and sheeting bleaching. Then again with goods sold in the white, it is customary to weave coloured headings or markings, and as these have to be preserved, to do so will cause some slight alteration of the details of the bleach with this object. On all these points it is difficult to lay down general rules because of the very varying feature of the conditions which are met with by the market bleacher.

The resin boil may be omitted, only two lye boils being required, and these need not be so long or of such a searching [Pg 40] character as the corresponding boils of the madder bleach. Outlines of two or three such processes, which are now in use in bleach works, will serve to show the general routine of a market bleach. The proportions given are calculated for 10,000 lb. of goods:—

(1) Lime boil, using 500 lb. of lime, and giving a twelve-hours' boil.

(2) Grey sour, hydrochloric acid of 2° Tw., then wash well.

(3) Lye boil, 100 lb. caustic soda, 70 per cent. solid, ten to twelve hours' boil; wash.

(4) Second lye boil, 100 lb., 58 per cent. soda ash, twelve-hours' boil.

(5) Chemic, bleaching powder liquor at 1° Tw., boil for three hours; wash.

(6) White sour, sulphuric acid at 2° Tw.; wash well.

The length of boil with the lime and lyes will depend upon the quality of the goods, heavy goods will require from two to three hours longer than will light goods, such as cambrics, the time given above being that for heavy goods, sheetings, etc.

Another process is the following: —

(1) Lime boil, using 480 lb. lime, and boiling for ten hours.

(2) Grey sour, hydrochloric acid at 2° Tw.; wash.

(3) Lye boil, 300 lb. soda ash, 58 per cent.; 50 lb. caustic soda, 70 per cent., and 30 lb. soft soap, giving ten hours' boil; wash.

(4) Chemic as above.

(5) White sour as above; wash well.

A smaller quantity of lime is used here, but on the other hand the lye boil is a stronger one. This process gives good results. Some bleachers do not use lime in their market bleaches, but give two lye boils, in which case the process becomes: [Pg 41] —

(1) Lye boil, using 140 lb. caustic soda, of 70 per cent., giving ten hours' boil and washing well.

(2) Second lye boil, using 120 lb. soda ash, 58 per cent., and giving ten hours' boil; wash.

(3) Chemic as above.

(4) White sour as above; wash well.

Light fabrics, such as laces, lace curtains, muslins, etc., cannot stand the strain of the continuous process, and they are therefore subjected to a different bleaching process, which varies much at different bleach works. One method is to lime by steeping for an hour in a weak lime liquor, using about 2 lb. of lime to 100 lb. of goods; they are then boiled in the kier for eight hours, after which they are washed. This washing is done in what are called dash wheels, large hollow wheels, the interior of each being divided into four compartments. Into these the goods are put, and the wheel is caused to revolve, while at the same time a current of water flows with some force into the interior of the wheel and washes the goods.

The wheels do their work well, and the action being gentle the finest fabrics can be washed without fear of any damage. It is necessary that the speed at which they are driven should be such that as the wheel revolves the goods are thrown from side to side of each compartment; if the speed be too slow they will simply slide down, and then they do not get properly washed; on the other hand, if the speed be too great then centrifugal action comes into play and the goods remain in a stationary position in the wheels with the same result. As to the amount of washing, it should be as before. After this washing they are boiled again in the kier with soda ash, using about 8 lb. ash for 100 lb. goods and giving seven hours' boil, which, after washing, is followed by a second boil with 5 lb. ash and 4 lb. soft soap for each 100 lb. of goods, giving eight hours' boil. They are then washed and [Pg 42] entered into the chemic. The chemicing is done in stone cisterns, which are fitted with false bottoms, on which the goods can rest, and which allow any insoluble particles of bleaching powder to settle out and prevent them from getting on the goods. The liquor is used at the strength of about ½° Tw., and the goods are allowed to steep about two hours; they are then placed in a hydro-extractor and the surplus chemic is whizzed out, after which they are steeped in sour of hydrochloric acid at 1° Tw., kept in a stone cistern, the goods being allowed to steep for two hours. Next they are washed, well whizzed, passed through a blueing water, whizzed again, and dried. The remarks made when describing similar operations of the madder bleach as to the action, testing, etc., of the chemicals, are equally applicable here.

Another plan of bleaching fine fabrics is shown in outline in the following scheme: —

(1) Wash; boil in water for two hours.

(2) Boil in soda for five hours, using 80 lb. soda ash of 58 per cent., and 30 lb. soft soap for 1,000 lb. of goods.

(3) Second soda boil, using from 40 lb. to 50 lb. soda ash, and 15 lb. to 20 lb. soft soap, giving four hours' boil; after each soda boil the goods should be washed.

(4) Chemic, using bleaching powder liquor at ½° Tw., allowing them to steep for two hours, then wash and whiz.

(5) White sour, using hydrochloric acid at 2° Tw., steeping two hours; wash.

A further extension of the same process is sometimes given for the best goods, which consists, after the above, in giving:—

(6) A third soda boil, using 25 lb. to 30 lb. soda ash and 15 lb. to 20 lb. soft soap, giving one hour's boil; washing.

(7) Chemic as before.

(8) Sour as before, after which the goods are well washed.

In the bleaching of Nottingham lace curtains for the soda [Pg 43] boils there is used what is called the "dolly," which consists of a large round wooden tub about 5 feet to 6 feet in diameter and about 2 feet 6 inches to 3 feet deep; this is made to revolve slowly at about one revolution per minute. Above the tub on a strong frame are arranged four stampers or beaters, which are caused to rise and fall by means of cams. The goods are placed in the tub with the scouring liquors and the dolly is set in motion, the beaters force the liquor into the goods, and the revolution of the tub causes the beaters to work on a fresh portion of the goods at every fall.

This is rather an old-fashioned form of machine, and is being replaced by more modern forms of boiling kiers. In bleaching certain kinds of muslins in which the warp threads are double, and in the case of lace curtains, it is necessary to endeavour to keep the threads as open and prominent as possible. This cannot be done with the continuous process, which puts a strain on the threads and thus effaces their individuality. To avoid this the fabrics have to be dealt with in bundles or lumps, and on these no strain is put, therefore every thread retains its individuality. The process above described is applicable.

Yarn Bleaching.—Yarn is supplied to the bleacher in two forms: (1) warps in which the length of the threads may vary from as little as 50 to as much as 5,000 yards; these can be dealt with in much the same manner as a piece of cloth, that is, a continuous system can be adopted; (2) hanks, which are too well known to require description. Sometimes yarn is bleached in the form of cops, but as the results of cop bleaching are not very satisfactory it is done as little as possible.

Warp Bleaching.—The warp, if very long, is doubled two, three or four times upon itself, so as to reduce its length. Care should be taken that the ends of the warp are tied together to prevent any chance of entangling, which would [Pg 44] very likely happen if the ends were left loose to float about. As a rule, warps are not limed, but the adoption of the liming would assist the bleaching. In outline warp bleaching consists of the following operations:—

(1) Lye boil, using 30 lb. caustic soda, 70 per cent., and 50 lb. soda ash, 58 per cent., giving six hours' boil, and washing.

(2) Sweeting, boil with 80 lb. soda ash, 58 per cent., for two hours.

(3) Washing.

(4) Chemicing, bleaching powder liquor at 1° Tw., washing.

(5) Sour, sulphuric acid at 2° Tw,. washing well.

(6) Hydro extracting and drying.

About 2,000 to 3,000 lb. of warps are usually treated at one time.

The machinery used may be the same as that used in the cloth bleach, and each operation may be conducted in the same manner. In some warp bleachworks, while the kiers are made in the same way, the other machines are made differently. The chemicing and souring is done in strong cisterns provided with a false bottom; in these the warps are allowed to remain for about two hours. A more complicated form of chemicing cistern is also in use. This is made of stone, and is provided with a false bottom. Above is a tank or sieve, as it is called, having a perforated bottom through which the liquor flows on the warp in the cistern below.

Under the chemicing cistern is a tank into which the liquor flows, and from which it is pumped up into the sieve above. A circulation of liquor is thus kept up during the whole of the operation. Owing to the action of the chemic or acid on the metal work of the pump there is great wear and tear of the latter, necessitating frequent repairs. This is a defect in this form of chemicing machine. For drying the warps a hydro-extractor is first used to get the surplus [Pg 45] liquor from the goods. This machine is now well known, and is in use in every bleachworks, where it is familiarly known as the "whiz," and the operation is generally called whizzing. Hydro-extractors are described under the head of "Dyeing Machinery".

The actual drying of the warps is done over the "tins" as they are called. These are a number of large cylinders measuring about 20 inches in diameter and about—for warp drying—5 feet long. Usually they are arranged vertically in two tiers, each tier consisting of about five cylinders, not arranged directly one above another but in a zig-zag manner, the centres of the first, third and fifth being in one line, and the centres of the others in another line. The cylinders are made to revolve by suitable driving mechanism, and into them is sent steam at about 5 lb. to 10 lb. pressure, which heats up the cylinders, whereby the warp passing over them is dried. This drying may be partial or complete, being regulated by the speed at which the warps pass over the cylinders and by the quantity of steam passed into the same. The quicker the speed and the smaller the amount of steam, the less the warps are dried; while, on the other hand, the slower the speed and the larger the amount and greater the pressure of the steam, the quicker and more thoroughly are the warps dried. As there is a great deal of water formed in the cylinders by the condensation of the steam, means are always provided for carrying off this water, as its retention in the cylinders often leads to serious results and damage to the machine.

Hank Bleaching.—So far as the chemical part of hank bleaching is concerned it does not differ from that of warp bleaching; the same operations and proportions of chemicals may be used and in the same order, but there is some difference in the machinery which is used. The hanks may be manipulated in two ways: they may be either kept in separate hanks, which is the method mostly in vogue in modern bleach- [Pg 46] houses, or they may be linked together in the form of a chain. In the latter case the operations and the machinery may be the same as used in the madder bleach, with a few unimportant minor differences. In the final washing the dumping machine is used, which consists of two wooden bowls set over a wooden trough containing the wash waters. The top bowl is covered with a thick layer of rope and merely rests on the bottom bowl by its own weight, and is driven by friction from the latter. The chain of hanks passing through between the two bowls has the surplus liquor squeezed out of it, and as there is considerable increase in the thickness at the points of linkage between the hanks, when these pass through the bowls they lift up the top bowl, which, when the thick places have passed through, falls down with a sudden bump upon the thin places, and this bumping drives out all the surplus liquor and drives the liquor itself into the very centre of the hanks, which is sometimes an advantage.

In modern bleach-houses the chain form is gradually giving place to the method of bleaching separate hanks, partly because so many improvements have been made in the hank-bleaching machinery of late years, which enables bleachers to handle the yarn in the form of separate hanks better than they could do formerly; and as bleaching in separate hanks means that the cotton is kept in a more open form, and is thus more easily penetrated by the various liquors which are used, it follows that the bleach will be better and more thorough, which is what the bleacher aims at. At the same time weaker liquors or, what is the same thing, less material can be used, which means a saving in the cost of the process. For bleaching yarn in the hank the following process may be followed with good results: —

(1) Lye boil, using 1,000 lb. yarn, 40 lb. caustic soda of 70 per cent., and 50 lb. of soda ash of 58 per cent., giving five to six hours' boil at low pressure. [Pg 47]

(2) Wash through washing machine.

(3) Second lye boil, using 40 lb. soda ash of 58 per cent., and giving two to three hours' boil, wash again through a washing machine.

(4) Chemic as in warp bleaching.

(5) Sour as in warp bleaching.

(6) Wash well.

(7) Hydro extract and dry.

Sometimes, if the yarn is to be sold in thread form, before the last operation it passes through another, *viz.*, treating with soap and blue liquors, which will be dealt with presently.

The lye boils are done in the ordinary kiers, and do not call for further notice, except that in filling the goods into the kiers care should be taken that while sufficiently loose to permit of the alkaline liquors penetrating through the hanks properly, yet they should be so packed that they will not float about and thus become entangled and damaged.

The washing is nowadays done in a special form of washing machine, designed to wash the hanks quickly and well with as little expenditure of labour and washing liquor as possible. There are now several makes of these washing machines on the market, most of them do their work well, and it is difficult to say which is the best. Some machines are made to wash only one bundle at once, while others will do several bundles. Generally the principle on which they are constructed is the same in all, a trough containing the ash liquor, over which is suspended a revolving reel or bobbin, usually made of wood or enamelled iron, the bobbin being polygonal in form so that it will overcome readily any resistance the yarn may offer and carry the hank round as it revolves. The hank dips into the wash liquor in the trough, and as it is drawn through by the revolution of the bobbin it is washed very effectually. The moving of the hank opens [Pg 48] out the threads, and thus the wash liquor thoroughly penetrates to every part of the hank, so that a few minutes' run through this machine thoroughly washes the yarn. A constant stream of clean water is passed through the trough. This machine may also be used for soaping and sizing the hanks if required. By extending the trough in a horizontal direction, and increasing the number of reels or bobbins, the quantity of material that can be washed at one time can be extended, although not to an indefinite extent. The workman can start at one end of the machine and fill all the bobbins with yarn, by the time he has finished this the first bobbinful will have been washed sufficiently and can be taken off and replaced with another quantity of yarn, and thus one by one the bobbins may be emptied and refilled, which means that a considerable amount of material can be got through in the course of a day. To avoid the labour of walking to and fro to fill and refill the bobbins, washing machines are constructed in which the trough is made in a circular form. The bobbins are placed at the ends of radial arms which are caused to revolve round over the trough, the workman is stationed constantly at one part of the circle, and as the arms pass him in their motion round the trough he takes off the washed hanks and puts on the unwashed hanks. By this machine he is saved a very considerable amount of labour, and is able to do his work in a more convenient manner. The yarn is well washed in such a machine. The trough may be entire or it may be divided into a number of compartments, each of which may contain a different kind of wash liquor if necessary. Of course it goes almost without saying that in all these machines the liquors in them may be heated up by means of steam pipes if required.

The chemicing and souring of the hanks does not call for special mention, beyond the fact that these operations are done in the same manner as warp bleaching. In Fig. 5 is [Pg 49] shown Mather & Platt's yarn-bleaching kier, which is designed to bleach cotton yarn, either in hanks or in the warp forms, without removing it from the vessel into which it is first placed. The process is as follows: The hot alkali solution is circulated by means of a distributing pipe through the action of an injector or centrifugal pump to scour the yarn; then water is circulated by means of a centrifugal pump for washing. The chemic and sour liquors are circulated also by means of pumps, so that without the slightest disturbance to the yarn it is quickly and economically bleached.

FIG. 5.—Mather & Platt's Yarn-bleaching Kier.

[Pg 50]

STAINS AND DAMAGES IN BLEACHED GOODS.

Some of the stains in bleached goods which are met are beyond the control of the bleacher to avoid, while others are due to various defects in the process. Now the subject of stains can only be dealt with in a very general way, because of the varying manner in which they arise. The recognition of the particular way in which the stains have been formed is sometimes difficult to discover. First, there are iron stains, which are the most common kind of stains that a bleacher is troubled with. These generally make their appearance in the form of red spots of greater or less extent. As a rule they are not visible before the pieces are fully bleached. Their origin is varied.

Sometimes they arise from the machinery; if the kiers are not kept thoroughly whitewashed out, there is a great liability to produce iron stains. Every other machine which is used in the process is made of iron, and should be kept free from rust, or the chances of stains are considerably increased. The water used in the bleaching must be free from iron. A small trace will not make much difference, but some waters contain a great deal of iron, so much so that they are absolutely unusable for bleaching purposes. Iron stains are often due to a very curious cause: the dropping of the oil used in the spinning or weaving machinery on to the cotton during the process of manufacture. This oil is often charged with iron derived from the wear and tear of the machinery, and which often gets fixed in the form of red spots of oxide on the fibre. Iron stains cannot readily be extracted.

Oil stains are also common. These take the form of bright yellow stains in various shapes, sometimes extending along the piece in streaks, at other times in patches in various places about the piece. Generally these oil stains do not make their appearance as soon as the piece is bleached, [Pg 51] and often the bleacher sends out his goods quite white and apparently all right, and yet soon afterwards comes a complaint that the goods are stained yellow. One cause of these yellow oil stains can be traced to the use of paraffin wax in the sizing of the warps. In this case the stains are more or less streaky in form, and extend along the length of the piece. They are due to the fact that paraffin wax is not saponifiable by the action of the alkalies used in the process, and is therefore not extracted. When the goods are chemiced the chlorine acts upon the paraffin and forms chlorine compounds, which are acted upon by light, and turn yellow by exposure to that agent and to the atmosphere. Paraffin, when used for the sizing of warps, may sometimes be completely extracted from the fabric, but this depends upon the proportion of tallow or other fat which is used in the composition of the sizing grease. If the paraffin is only present in small quantities, and the grease well mixed, then it may be possible to extract all the paraffin out of the fabric during the bleaching process, but if the paraffin is in large proportion, or the grease not well mixed, it is scarcely possible to extract it all out, and stains must be the result. These stains can hardly be considered the fault of the bleacher, but are due to the manufacturer of the cloth using cheap sizing compositions on his warps. There are no means which can be adopted before bleaching to ascertain whether paraffin exists in the cloth. If found to be present, the remedy which is the easiest practically is to saturate the cloth with a little olive oil, or better, pale oleic acid. Allow the fatty matter to soak well in, and then boil the goods in a little caustic soda. Another cause of oil stains is the use of mineral oils in the lubrication of cotton machinery. These mineral oils partake of the nature of paraffin in their properties, and therefore they are unsaponifiable [Pg 52] by the action of alkalies. Like paraffin wax, they resist the bleaching process, and much in the same manner produce stains. Oil stains show themselves in various forms—sometimes as spots. These may be due to the splashing of oil from the spindles during the process of spinning, or they may be in patches of a comparatively large size over the pieces.

These are perhaps due to the oil dropping on to the piece during the process of weaving when in the loom. The oils used for the lubrication of spinning and weaving machinery should contain a fair proportion of some fatty oil, such as olive or rape or cocoanut oil. Not less than 10 per cent. should be used. More would be better, but the cost of course would be greater and oil is an item with spinners and manufacturers.

Stains are occasionally due to other causes rather too numerous to be dealt with in detail, and sometimes these stains only appear once in a lifetime, and often do not make their appearance during the bleaching process, but only in after dyeing or calico printing processes in curious ways the causes of which are very baffling to find out. [Pg 53]

CHAPTER III.

DYEING MACHINERY AND DYEING MANIPULATIONS.

Cotton is dyed in a variety of forms: raw, loose cotton, partly manufactured fibre in the form of slubbing or sliver, spun fibres or yarns wound in cop or bobbin forms, in hanks or skeins and in warps, and lastly in the form of woven pieces. These different forms necessitate the employment of different forms of machinery and different modes of handling; it is evident to the least unobservant that it would be quite impossible to subject slubbing or sliver to the same treatment as yarn or cloth, otherwise the slubbing would be destroyed and rendered valueless.

In the early days all dyeing was done by hand in the simplest possible contrivances, but during the last quarter of a century there has been a great development in the quantity of dyeing that has been done, and this has really necessitated the application of machinery, for hand work could not possibly cope with the amount of dyeing now done. Consequently there has been devised during the past two decades a great variety of machines for dyeing every description of textile fabrics, some have not been found a practical success for a variety of reasons and have gone out of use, others have been successful and are in use in dye-works.

HAND DYEING.

Dyeing by hand is carried on in the simplest possible appliances; much depends upon whether the work can be [Pg 54] done at the ordinary temperature or at the boil. Figs. 6 and 7 show respectively a rectangular vat and a round tub much in use in dye-houses. These are made of wood, but

FIG. 6. — Rectangular Dye-tank.

7.—Round Dye-tub.

copper dye-vats are also made. These may be used for all kinds of material, loose fibre, yarns or cloth. In the case of loose fibre this is stirred about either with poles or with rakes, care being taken to turn every part over and over and open out the masses of fibre as much as possible in order to [Pg 55] avoid matting or clotting together. In the case of yarns or skeins, these are hung on sticks resting on the edges of the tub or vat. These sticks are best made of hickory, but ash or beech or any hard wood that can be worked smooth and which does not swell much when treated with water may be used. The usual method of working is to hang the skein on the stick, spreading it out as much as possible, then immerse the yarn in the liquor, lift it up and down two or three times to fully wet out the yarn, then turn the yarn over on the stick and repeat the dipping processes, then allow to steep in the dye-liquor. This is done with the batch of yarn that is to be dyed at a time. When all the yarn has been entered into the dye-bath, the first stickful is lifted out, the yarn turned over and re-entered in the dye-liquor, this operation is carried out with all the sticks of yarn until the cotton has become dyed of the required depth. In the case of long rectangular vats it is customary for two men, one on each side of the vat to turn the yarns, each man taking charge of the yarn which is nearest to him. The turning over one lot of yarn is technically called "one turn" and the dyer often gives "three turns" or "four turns" as may be required.

Woven goods may be dyed in the tub or vat, the pieces being drawn in and out by poles, but the results are not altogether satisfactory and it is preferable to use machines for dyeing piece goods.

Plain tubs or vats, such as those shown in Figs. 6 and 7, are used for dyeing, and otherwise treating goods in the cold, or at a lukewarm heat, when the supply of hot water can be drawn from a separate boiler. When, however, it is necessary to work at the boil, then the vat must be fitted with a steam coil. This is best laid along the bottom in a serpentine form. Above the pipe should be an open lattice-work bottom, which, while it permits the free circulation [Pg 56]

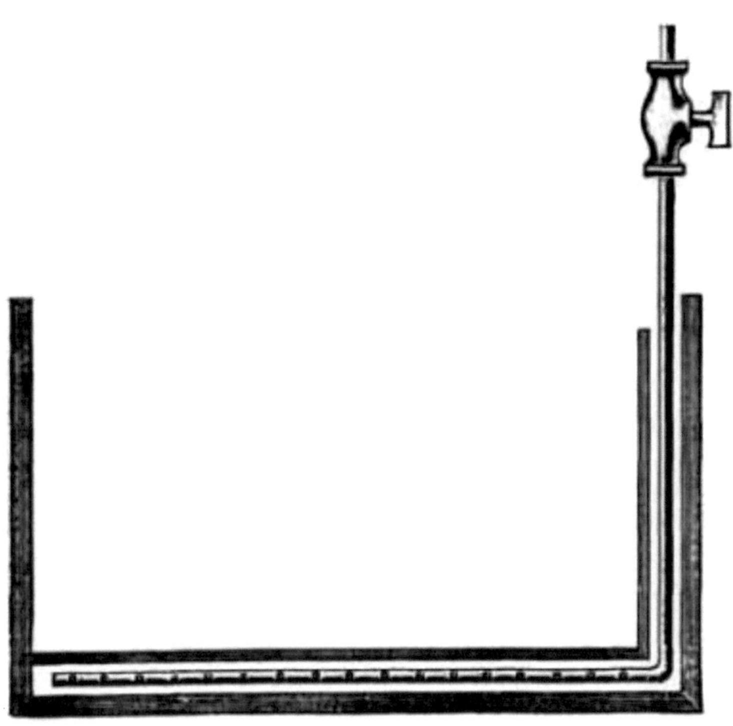

FIG. 8.—Section of Dye-vat.

of boiling water in the vat, prevents the material being dyed from coming in contact with the steam pipe. This is important if uniform shades are to be dyed, for any excessive heating of any portion of the bath leads to stains being produced on the material in that part of the bath. Fig. 8 shows a vat fitted with a steam pipe. That portion of the steam pipe which passes down at the end of the vat is in a small compartment boxed off from the main body of the vat, so that no part of the material which is being dyed can come in contact with it. A closed steam coil will, on the whole, give the best results, as then no weakening of the dye-liquor can take place through dilution by the condensation of the steam. Many dye-vats are, however, fitted with perforated, or, as they are called, open steam coils, in which case there is, perhaps, better circulation of the liquor in the dye-vat, but as some of the steam must [Pg 57] condense, there is a little dilution of the dye-liquor in the vat.

DYEING MACHINES.

Dye tubs and vats, such as those described above, have been largely superseded by machines in which the handling, or working of the materials being dyed is effected by mechanical means. There have been a large number of dyeing machines invented, some of these have not been found to be very practical, and so they have gone out of use. Space will not admit of a detailed account of every kind of machine, but only of those which are in constant use in dye-works.

Dyeing Loose, or Raw Cotton.—Few machines have been designed for this purpose, and about the only successful one is:

Delahunty's Dyeing Machine.—This is illustrated in Fig. 9. It consists of a drum made of lattice work, which can revolve inside an outer wooden casing. The interior of the revolving drum is fitted with hooks or fingers, whose action is to keep the material open. One segment of the drum is made to open, so that the loose cotton or wool to be dyed can be inserted. By suitable gearing the drum can be revolved; and the dye-liquor, which is in the lower half of the wooden casing, penetrates through the lattice work of the drum, and dyes the material contained in it. The construction of the machine is well shown in the drawing, while the mode of working is obvious from it and the description just given. The machine is very successful, and well adapted for dyeing loose, or raw wool and cotton. The material may be scoured, bleached, dyed, or otherwise treated in this machine.

The Obermaier machine, presently to be described, may also be used for dyeing loose cotton or wool. [Pg 58]

DYEING, SLUBBING, SLIVER OR CARDED COTTON AND WOOL.

FIG. 9.—Delahunty's Dyeing Machine.

It is found in practice that the dyeing of loose wool or cotton is not altogether satisfactory—the impurities they naturally contain interfere with the purity of the shade they will take. Then again the dyes and mordants used in dyeing them are found to have some action on the wire of the carding engine through which they are passed; at any rate a card does not last as long when working dyed cotton or wool as when used on undyed cotton or wool fibres. Yet for the production of certain fancy yarns for weaving some special classes of fabrics, it is desirable to dye the cotton or wool [Pg 59]

FIG. 10.—Obermaier Dyeing Machine.

before it is spun into thread. The best plan is undoubtedly to dye the fibre after it has been carded and partly spun into what is known as slubbing or sliver. All the impurities have been removed, the cotton fibres are laid, straight, and so it becomes much easier to dye. On the other hand, as it is necessary to keep the sliver or slubbing straight and level, no working about in the dye-liquors can be allowed to take place, and so such must be dyed in specially constructed machines, and one of the best of these is the Obermaier dyeing machine which is illustrated in Fig. 10. The Obermaier apparatus consists of a dye vat A. In this is placed a cage consisting of an inner perforated metal cylinder C, and an outer perforated metal cylinder D, between these two is placed the material to be dyed. C is in contact with the suction end of a centrifugal pump P, the delivery end of which discharges into the dye-vat A. The working of the machine is as follows: The slubbing or sliver is placed in the [Pg 60] space between C and D rather tightly so that it will not move about. Then the inner cage is placed in the dye-vat as shown. The vat is filled with the dye-liquor which can be heated up by a steam pipe. The pump is set in motion, the dye liquor is drawn from A to C, and, in so doing, passes through the material packed in B and dyes it. The circulation of the liquor is carried on as long as experience shows to be necessary. The dye-liquor is run off, hot water is run in to wash the dyed material, and the pump is kept running for some time to ensure thorough rinsing; then the water is run off, and by keeping the pump running and air going through a certain amount of drying can be effected. This machine works very well, and, with a little experience, constant results can be obtained. The slubbing or sliver may be scoured, bleached, rinsed, dyed, washed, soaped, or otherwise treated without removing it from the machine, which is a most decided advantage.

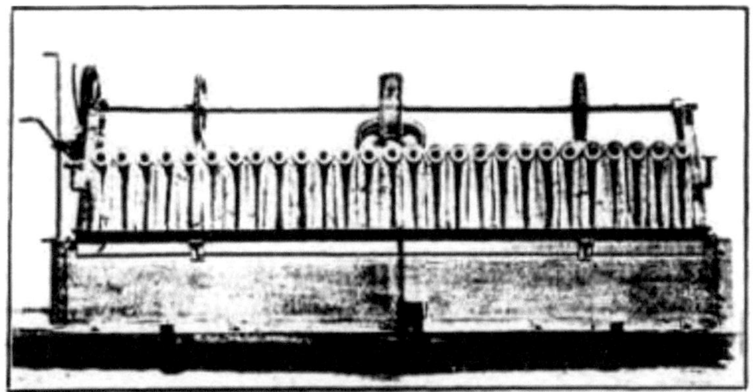

FIG. 11.—Holliday's Yarn-dyeing Machine.

Holliday's Yarn-dyeing Machine.—In Fig. 11 is given an illustration of a machine for dyeing yarn in the hank form made by Messrs. Read Holliday & Sons, of Huddersfield. The illustration gives a very good idea of the [Pg 61] machine. It consists of a wooden dye-vat which can be heated by steam pipes in the usual way. Extending over the vat are a number of reels or bobbins; these are best made of wood or enamelled iron; these reels are in connection with suitable gearing so that they can be revolved. There is also an arrangement by means of which the reels can be lifted bodily in and out of the dye-vat for the purpose of taking on and off, "doffing," the hanks of yarn for the reels. A reel will hold about two pounds of yarn. The working of the machine is simple. The vat is filled with the requisite dye-liquor. The reels, which are lifted out of the vat, are then charged with the yarn, which has been previously wetted out. They are then set in revolution and dropped into the dye-vat and kept there until it is seen that the yarn has acquired the desired shade. The reels are lifted out and the hanks removed, when the machine is ready for another lot of yarn.

There are several makes of hank-dyeing machines of this type, and as a rule they work very well. The only source of trouble is a slight tendency for the yarn on one reel if hung loosely of becoming entangled with the yarn on one of the other reels. This is to some extent obviated by hanging in the bottom of the hank a roller which acts as a weight and keeps the yarn stretched and so prevents it flying about.

To some makes of these machines a hank wringer is attached.

Klauder-Weldon Hank-dyeing Machine.—This is illustrated in Fig. 12, which shows the latest form. This machine consists of a half-cylindrical dye-vat built of wood. On a central axis is built two discs or rod carriers which can revolve in the dye-vat, the revolution being given by suitable gearing, which is shown at the side of the machine. On the outer edge of the discs are clips for carrying rods, on which one end of the hanks of yarn is hung, while the other [Pg 62]

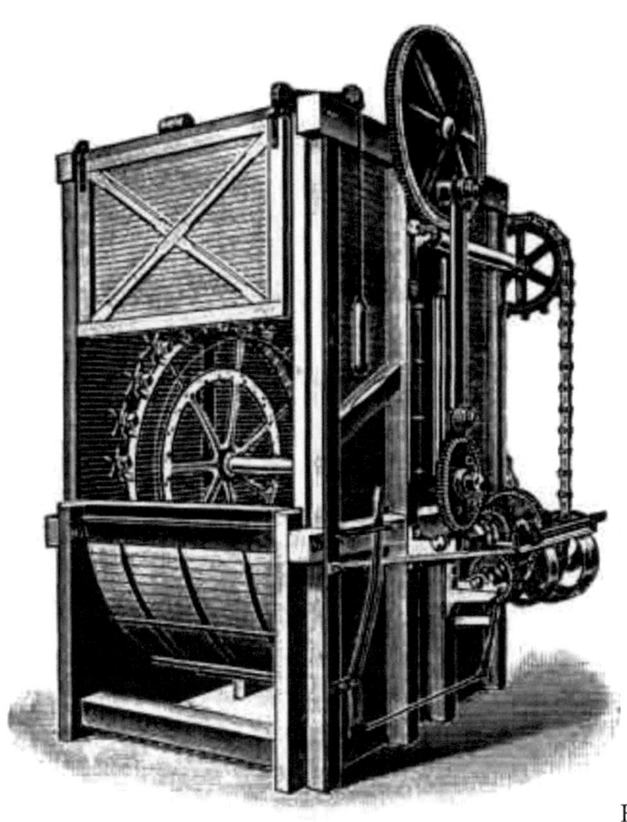

FIG. 12. —

Klauder-Weldon Dyeing Machine.

end is placed on a similar rod carried near the axle. The revolution of the discs carries the yarn through the dye-liquor contained in the lower semi-cylindrical part of the machine previously alluded to. At a certain point, every revolution of the discs, the rods carrying the yarns are turned a little; this causes the yarn to move on the rods, and this motion helps to bring about greater evenness of dyeing. The most modern form of this machine is provided with an arrangement by means of which the whole batch of yarn can be lifted out of the dye-liquor. Arrangements are made by which from time [Pg 63] to time fresh quantities of dye can be added if required to bring up the dyed yarn to any desired shade. This machine works well and gives good results. Beyond the necessary labour in charging and discharging, and a little attention from time to time, as the operation proceeds, to see if the dyeing is coming up to shade, the machine requires little attention.

Many other forms of hank-dyeing machines have been devised: there is Corron's, in which an ordinary rectangular dye-vat is used. Round this is a framework which carries a lifting and falling arrangement that travels to and fro along the vat. The hanks of yarn are hung on rods of a special construction designed to open them out in a manner as nearly approaching handwork as is possible. The machine works in this way: the lifting arrangement is at one end of the vat, the hanks are hung on the rods and placed in the vat. Then the lifter is set in motion and moves along the vat; as it does so it lifts up each rod full of yarn, turns it over, opening out the yarn in so doing, then it drops it again in the vat. When it has travelled to the end of the vat it returns, picking up the rods of yarn in so doing, and this motion is kept up until the dyeing is completed. This machine is very ingenious.

A type of machine which has been made by several makers consists of an ordinary rectangular dye-vat surrounded with a framework carrying a number of sets of endless chains, the links of which carry fingers. The hanks of yarn are hung on rods at one end of which is a tooth wheel that when in position fits into a rack on the side of the vat. The action of the machine is this: the hanks are hung on the rods and placed at the entrance end of the vat, by the moving of the chains it is carried along the vat and at the same time revolves, thus turning over the yarn which hangs in the dye-liquor; when it reaches the opposite end of the vat, the rod full of yarn is lifted out, carried upwards and then towards the other end of the vat when it is again [Pg 64] dropped into the dye-vat to go through the same cycle of movements which is continued until the yarn is properly dyed.

COP DYEING.

In weaving fancy-coloured fabrics the ordinary mode is to dye the yarn in the hank form, then those which have to be used for the weft are wound into the cop form for placing in the shuttles. The cop form is that in which the yarn leaves the spinning frame, and necessarily apart from the dyeing there is labour involved in reeling it into hanks and winding it back again into the cop form, not only so but there is necessarily some waste made in these operations. Many attempts have been made, with more or less success, to dye the yarn while in the cop form and so save the cost of the hanking and copping above referred to as well as the waste which occurs. Cops cannot be satisfactorily dyed by simple immersion in a boiling dye-bath, the outside becomes dyed but the central portions as often as not remain quite white, and there is a distinct grading of colour or shade throughout the cop, the outer portions being deeply dyed while the middle portion will only have a medium shade and the central portions either not being dyed at all or only faintly tinted, much depending on the firmness with which the cop has been wound. A soft, loosely wound cop is much more thoroughly dyed than a hard, tightly wound cop. This uneven dyeing of the cops is not satisfactory, and must be avoided if cop dyeing is to be a success. Many dyers have turned their attention to this question of dyeing yarn in the cop form, and many machines have been devised for the purpose; some of these have not been a success, but a few have been found to yield satisfactory results and proved in practice very successful.

In all machines for dyeing cops one principle has been [Pg 65] adopted—that of drawing or forcing the dye-liquor through the cop.

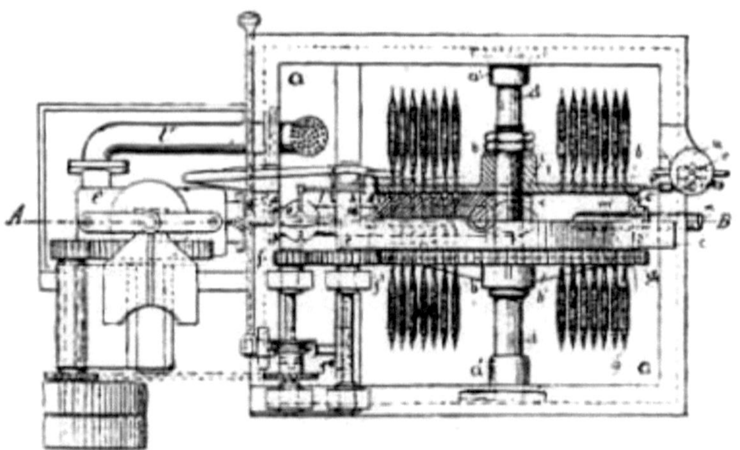

FIG. 13.—Graemiger Cop-dyeing Machine.

Graemiger's Cop-dyeing Machine.—This is shown in section in Figs. 13 and 14. Although simple in its work it is somewhat complex in its construction and difficult to describe. The machine consists of a dye-vat to hold the requisite dye-liquors. In the upper portions of this is an iron casting formed with four chambers, the two lower ones of which are immersed in the dye-liquor while the upper chambers are above it. The sides of this casting are formed of metal plates which fit tightly against the casting and form as nearly air- and water-tight joints with it as it is possible to make. These metal plates are on a spindle and can be rotated. They are perforated and made to carry spindles, on which are placed the cops to be dyed. The two lower chambers are in connection with a pump which draws the air from them and so creates a vacuum inside the chambers. To fill this, liquor from the dye-vat passes through the cops and into the chambers, and is in turn drawn through the pump and returned to the dye-vat. In this way there is a [Pg 66] continual circulation of dye-liquors from the vat through the cops, chambers and pump back to the vat again.

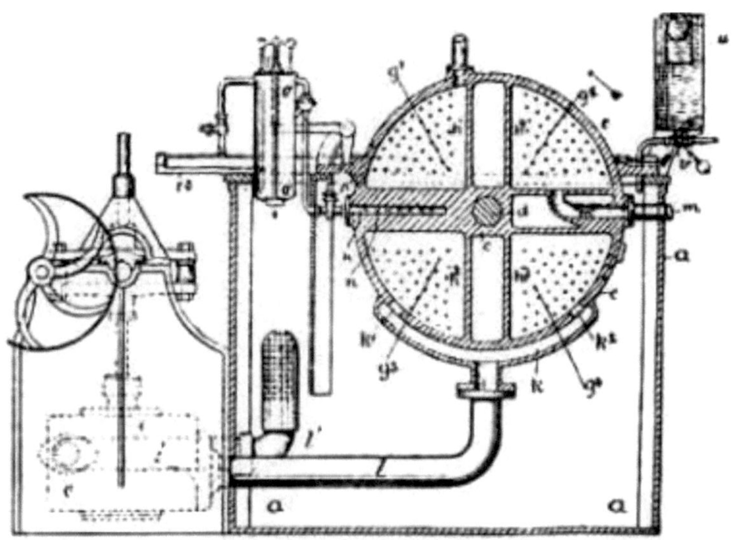

FIG. 14.—Graemiger Cop-dyeing Machine.

The left upper chamber is practically a blank chamber. Those portions of the cop carriers in contact with it are filled with cops, which are placed on perforated spindles; the discs are given a quarter revolution which brings the cops into the dye-liquor and in connection with the left lower chamber and are dyed. At the same time the section of the cop carriers now in contact with the left top chamber is filled with a new lot of cops, another quarter of a revolution is given to the cop carriers, which immerse the new lot of cops in the dye-liquor. The third quarter of the cop plates is filled with cops. A third movement of the cop plates now takes place; this brings the first lot of cops out of the dye-liquor and in contact with the right upper chamber, where the surplus liquor is drawn out of them and returned to the dye-vat. Another revolution brings the cops back to their first position, they are now removed and a new lot substituted. These [Pg 67] proceedings go on continuously. Although not quite free from defects the machine gives very good results, the cops being very uniformly dyed through.

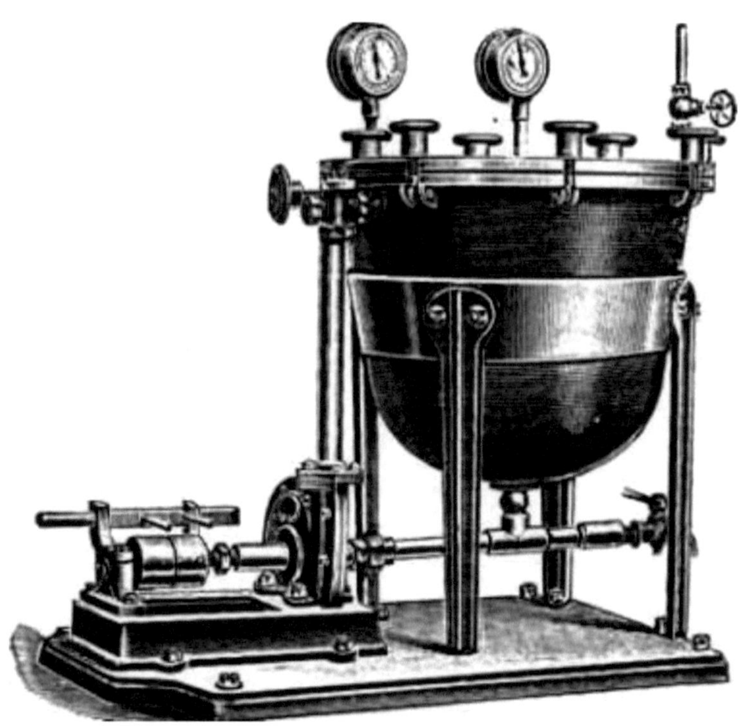

FIG. 15.—Beaumont's Cop-dyeing Machine.

Beaumont's Cop-dyeing Machine.—This is illustrated in Fig. 15. It consists of a copper hemispherical dye-vessel, which is provided with a tightly fitting lid, although this is not needed in all cases. The bottom of the vessel is in communication with the suction end of a centrifugal pump, while the delivery end of the pump is attached to the upper end of the dye-vessel, the action of the pump being to secure a constant circulation of dye-liquor from the bottom to the top of the dye-vessel. Arrangements are provided by a peculiar and ingenious contrivance fitted in one side of the dye-kettle for [Pg 68] introducing steam to heat the dye-liquor to any required degree. As in most forms of cop-dyeing machines, the cops are placed on perforated metal spindles. The cops and spindles are inserted in holes in a perforated metal plate, and over them is placed a thin metal plate, technically called the antifloater, whose object is to prevent the cops from becoming detached from the plate. This plate, full of

72

cops, is now placed in the dye-vessel and rests upon a flange which is provided for that purpose. When the cop plate is in position the dye-vessel is divided into two chambers—a lower chamber and an upper one, in the latter being the cops.

The pump draws liquor from the chamber under the cop plate and so creates a vacuum, which during the working of the machines ranges from 10 to 20 inches in degree. To supply this vacuum, dye-liquor is drawn from the upper chamber through the cops. The pump returns the liquor to the dye-vessel. A very rapid circulation of dye-liquor takes place, from 25 to 50 gallons per minute passing through the cops and pump. From five to ten minutes is sufficient to dye the cops. The machine is simple in its construction and gives good results, the cops being completely dyed through. One important consideration in cop dyeing is to be able to dye successive batches of cops to exactly the same shade, and this is quite possible with this machine.

Young & Crippin's Cop-dyeing Machine.—So far as simplicity of construction is concerned this lies between the two preceding machines. It consists of four parts with some accessory mechanism. There is first a dye-liquor storage tank at the base of the apparatus in which the liquor is kept stored and boiling (if necessary) ready for use, above this and at the front end is the dye-chamber, this communicates at its lower end by a pipe with the dye-liquor in the dye-vat. Then there is a large vacuum chamber, in which by means of an injector a vacuum can be formed, this directly com [Pg 69] municates with a liquor-receiving chamber which again in turn is in communication with the upper part of the dye-chamber. The cops are placed on perforated spindles as usual, and these on a perforated plate and are kept in place by a plate which is screwed down on them. The charged cop plate is placed in the dye-chamber on which a cover is placed and screwed down. By means of a lever the injector is set at work, a vacuum created in the vacuum and receiving chambers, the consequence being that dye-liquor is drawn from the vat through the cops in the dye-chamber into the receiving chamber. When a certain quantity of liquor has passed through, by a movement of a lever, the vacuum is destroyed, and the dye-liquor runs back into the dye-vat; these operations are repeated until from past experience of the working of the machine it is thought sufficient has passed through to dye the cops, when the dye-chamber is opened and the cops taken out. This machine works very well.

Mommer's Cop-dyeing Machine.—This is in use in several continental dye-works. The central portion of this machine is a rectangular dye-chamber, which can be hermetically closed by hinged doors, the cops are placed side by side on trays provided with perforated bottoms, the trays being placed one on the top of the other in the dye-chamber. From the top of the dye-chamber passes a pipe to a centrifugal pump, and a similar pipe passes from the bottom of the chamber to the pump. A separate vat contains the dye-liquor which is used. The pump forces the dye-liquor through the cops which take up the dye. Arrangements are provided by which the direction of the flow of the dye-liquor can be changed. This machine gives fairly good results, not perhaps equal to those with the machines previously described.

Warp-dyeing Machines.—Although many warps, especially for fancy fabrics, are prepared from yarns dyed in the hank [Pg 70] or cop form, yet it is found advantageous when a warp is of one colour, a self-colour as it is called, to form the warp from grey or white yarns and to dye it after warping. If the warp were so wound as to be able to go into a Obermaier dyeing machine, it would be possible to dye it in that machine, but generally warps are dyed in the open form and are passed through a dyeing vat, commonly called a warp-vat which is constructed as shown in Fig. 16. These warp-dyeing machines generally consist of a long rectangular wooden dye-vat, divided by two partitions into three compartments, each provided with steam pipes to heat up its contents; between the first and second and between the second and third compartments is fitted a pair of squeezing rollers, while the third compartment is fitted with a heavier pair of squeezing rollers. Motion is given to these rollers by suitable gearing, and they serve to draw the warp through the machine. Guide rollers are fitted in the compartment, and the warp being taken round these, it passes several times up and down and through the dye-liquors contained in the compartments. These warp-dyeing machines may be made of sufficient width to take one, two, three or more warps at one time as desired.

FIG. 16. — Warp-dyeing Machine.

The three compartments of the machine may contain different liquids or all the same liquid according as the nature of the shade to be dyed demands. The passage is done [Pg 71] slowly so as to give the warp time to absorb the liquors and take up the dye. When all the length of warp has been sent through, it is said to have been dyed "one end". Sometimes this will be enough, but often it is not, and so the warp is sent through again, given another end, and still again if the full shade has not been attained.

After being dyed in this machine the warp is sent through another one containing various wash liquors to finish the process.

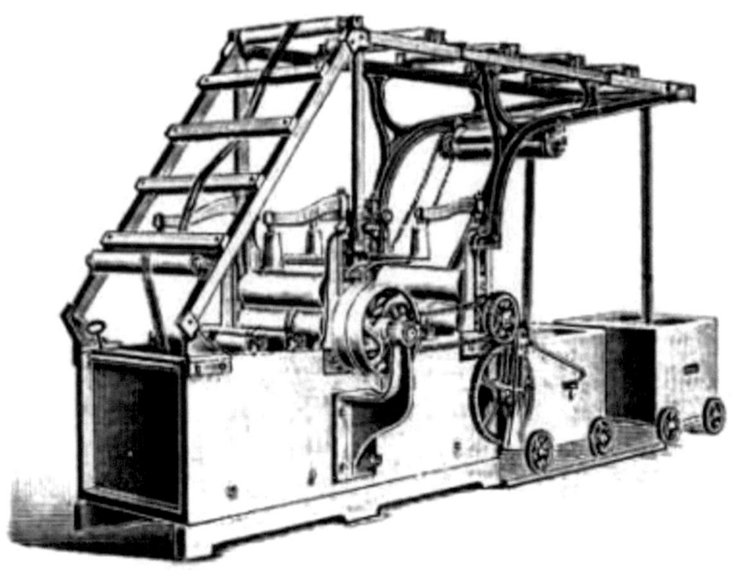

FIG. 17.—Warp-dyeing Machine.

Fig. 17 shows a warp-dyeing machine similar to, but a little more elaborate in construction than, the vats just described.

Piece-dyeing Machines.—Wherever it is possible it is far more preferable to dye textile fabrics in the form of woven pieces rather than in the yarn from which they are woven. During the process of weaving it is quite impossible to avoid the material getting dirty and somewhat greasy, and the operations of scouring necessary to remove this dirt and [Pg 72] grease has an impairing action on the colour if dyed yarns have been used in weaving it. This is avoided when the pieces are woven first and dyed afterwards, and this can always be done when the cloths are dyed in one colour only. Of course when the goods are fancy goods containing several colours they have to be woven from dyed yarns.

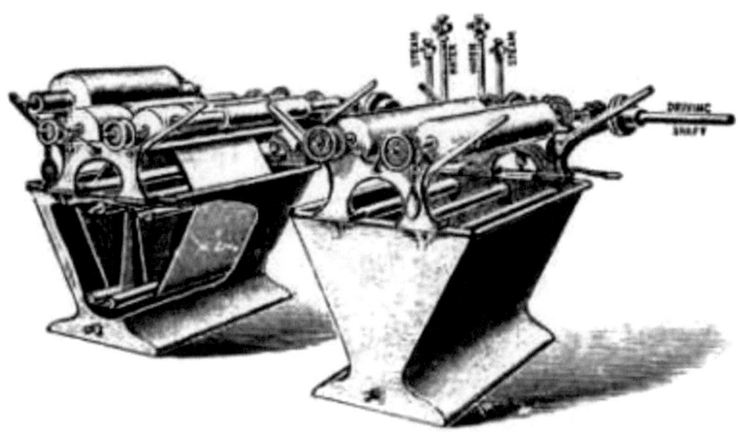

FIG. 18. — Dye-jiggers.

The most common form of machine in which pieces are dyed is The Jigger, commonly called the jig. This is shown in Figs. 18 and 19. It consists of a dye-vessel made sufficiently long to take the piece full width — wide at the top and narrow at the bottom. At the top at each side is placed a large winding roller on which the cloth is wound. At the bottom of the jig is placed a guide roller round which passes the cloth. In some makes of jigs (Fig. 19) there are two guide rollers at the bottom and one at the top, as shown in the illustration, so that the cloth passes several times through the dye-liquor. In working, the cloth is first wound on one of the rollers, then threaded through the guide rollers and attached to the other winding roller. When this is done dye-liquor is run into the jig, the gearing set in motion, and [Pg 73]

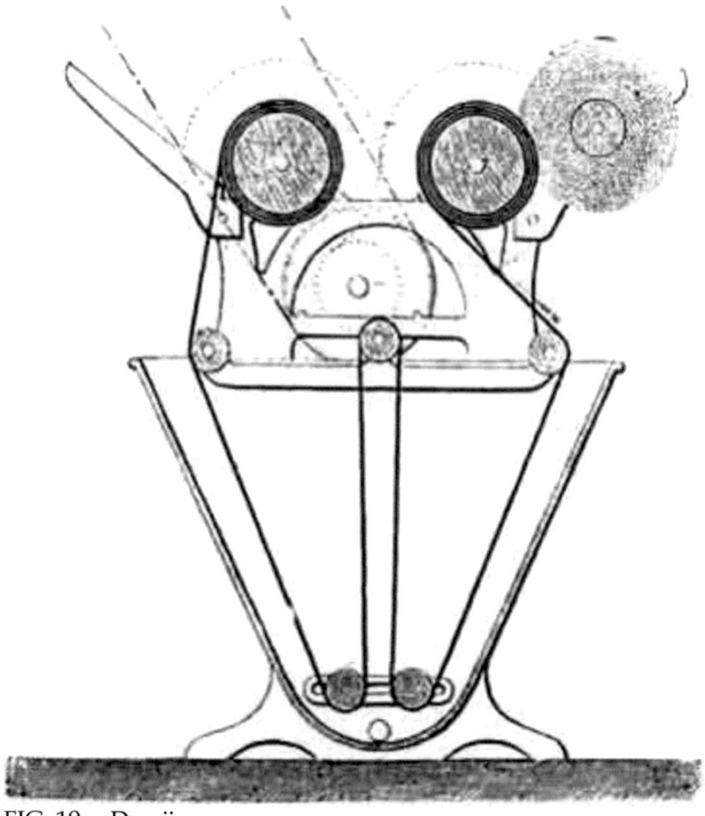

FIG. 19.—Dye-jigger.

the cloth wound from the full on to the empty roller. With the object of keeping the piece tight, a heavy press roller is arranged to bear on the cloth on the full roller. When all the cloth has passed from one roller to the other it is said to have been given "one end". The direction of motion is now changed, and the cloth sent in the opposite direction through the jig, and the piece has now received another "end". This alternation from one roller to the other is continued as long as is deemed necessary, much depending on the depth of colour which is being dyed—some [Pg 74] pale shades may only take two or three ends, deeper shades may take more. When dyeing wool with acid colours which are all absorbed from the dye-liquor, or the bath is exhausted, it is a good plan to run the pieces several more ends so as to ensure thorough fixation of the dye on the cloth.

It is not advisable in working these jigs to add the whole of the dye to the liquor at the commencement, but only a part of it; then when one end is given, another portion of the dye may be added; such portions being always in the form of solution. Adding dyes in powder form inevitably leads to the production of colour specks on the finished goods. The reason for thus adding the dye-stuff in portions is that with some dyes the affinity for the fibre is so great that if all were added at once it would all be absorbed before the cloth had been given one end; and, further, the cloth would be very deep at the front end, while it would shade off to no colour at the other end. By adding the dye in portions this difficulty is overcome and more level shades are obtained; it is met with in all cases of jigger dyeing, but it is most common in dyeing cotton or wool with basic dyes like magenta, auramine, methyl violet or brilliant green, and in dyeing wool with acid dyes like acid green, formyl violets, azo scarlet, or acid yellow.

Some attempts have been made to make jiggers automatic in their reversing action, but they have not been successful; owing to the greatly varying conditions of length of pieces, their thickness, etc., which have to be dyed, and it is next to impossible to make all allowances for such varying conditions.

The Jig Wince or Wince Dye Beck. — This dyeing machine is very largely used, particularly in the dyeing of woollen cloths. It is made by many makers, and varies somewhat in form accordingly. Figs. 20, 21 and 22, show three forms [Pg 75]

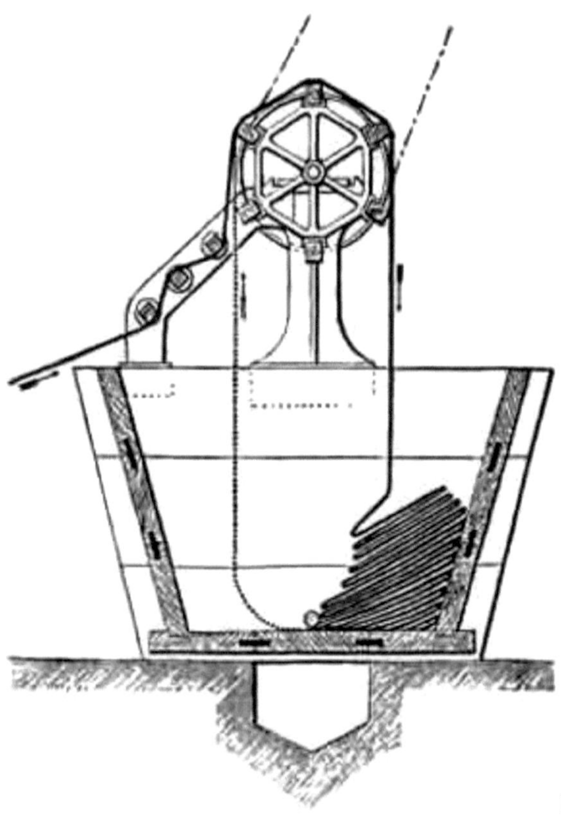

FIG. 20—Jig Wince.

by different makers. In any make the jig wince, or wince dye beck, consists of a large rectangular, or, in some cases, hemicylindrical dye-vat. Probably the best shape would be to have a vat with one straight side at the front, and one curved side at the back. In some a small guide roller is fitted at the bottom, under which the pieces to be dyed pass. Steam pipes are provided for heating the dye-liquors. The becks should be fitted with a false bottom made of wood, perforated with holes, or of wooden lattice work, and below

[Pg 76]

80

FIG. 21.—Cloth-dyeing Machine.

which the steam pipes are placed; the object being to prevent the pieces from coming in contact with the steam pipe, and so preventing the production of stains. Above the dye-vat, and towards the back, is the wince, a revolving skeleton wheel, which draws the pieces out of the dye-vat at the front, and delivers them into it again at the back. The construction of this wince is well shown in the drawings. The wince will take the pieces full breadth, but often they are somewhat folded, and so several pieces, four, five or six strings as they are called, can be dealt with at one time. In this case a guide rail is provided in the front part of the machine. In this rail are pegs which serve to keep the pieces of cloth separate, and so prevent entanglements. The pieces are stitched end to end so as to form an endless band. When running through the vat they fall down in folds at the back part of the [Pg 77] beck, and are drawn out of the bottom and up in the front. Each part thus remains for some time in the dye-liquor, during which it necessarily takes up the dye.

81

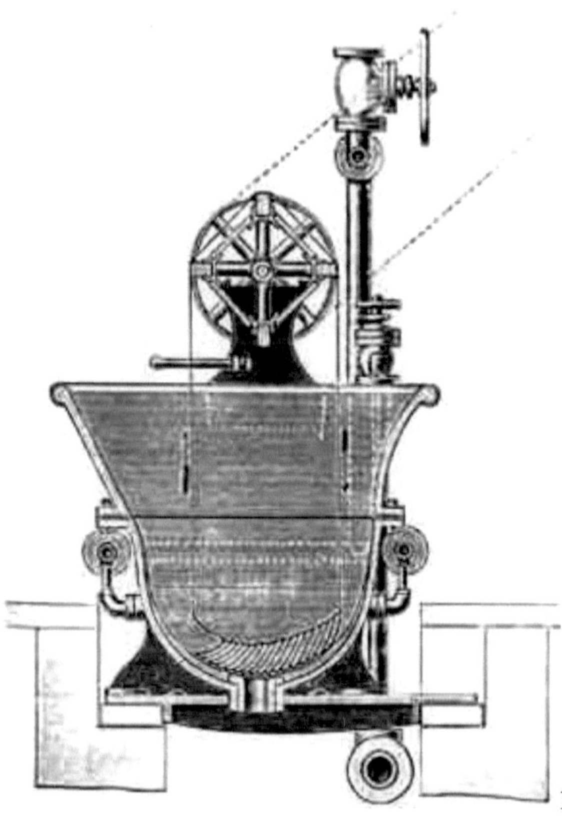

FIG. 22.—Dye Beck.

In the jig and wince dyeing machines the pieces necessarily are for a part of the time, longer in the case of the jigger than in that of the wince, out of the dye-liquor and [Pg 78] exposed to the air. In the case of some dyes, indigo especially, this is not desirable, and yet it is advisable to run the cloth open for some time in the liquor, so as to get it thoroughly impregnated with the dye-liquor, or to become dyed. This may be done on such a machine, as is shown in Fig. 24, page 79, but having all the guide rollers below the liquor, so that at no time is the piece out of the liquor, except, of course, when entering and leaving.

FIG. 23.—Holliday's Machine for Hawking Cloth.

The so-called hawking machines have also this object in view, and Fig. 23 is an illustration of Holliday's hawking machine, made by Messrs. Read Holliday & Sons, of Huddersfield. There is the dye-vat as usual; in this is suspended the drawing mechanism, whose construction is well shown in the drawing. This is a pair of rollers driven by suitable gearing, between which the cloth passes, and by which it is drawn through the machine. A small roller ensures the cloth properly leaving the large rollers; then there is a lattice-work arrangement over which the pieces are drawn. In actual work the whole of this arrangement is below the [Pg 79] surface of the dye-liquor in the vat. The piece to be dyed is threaded through the machine, the ends stitched together. Then the arrangement is lowered into the dye-vat and set into motion, whereby the cloth is drawn continuously in the open form through the dye-liquor, this being done as long as experience shows to be necessary. This hawking machine will be found useful in dyeing indigo on cotton or wool, or in dyeing cotton cloths with such dyes as Immedial blacks, Cross-dye blacks, Amidazol blacks, Vidal blacks, where it is necessary to keep the goods below the surface of the dye-liquor during the oper-ation.

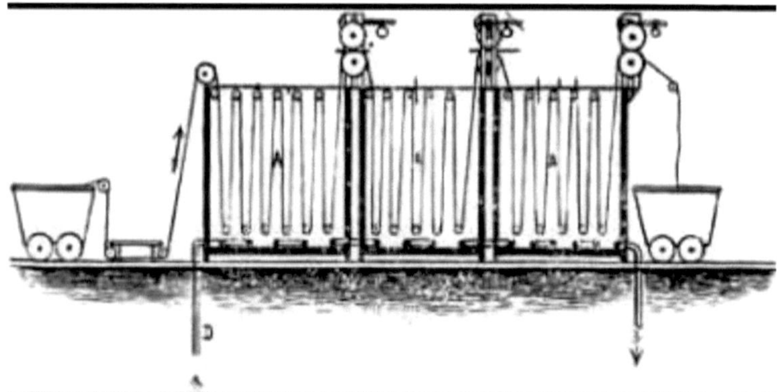

FIG. 24.—Continuous Dyeing Machine.

Fig. 24 shows a form of cloth-dyeing machine much used in the cotton trade. It consists of a number of compartments fitted with guide rollers at top and bottom, and round which the cloth is threaded, so that it passes up and down in the dye-liquor several times. Between each two compartments is a pair of squeezing rollers to press out all surplus liquors. All the compartments may be filled with the same dye-liquor, or with different dye-liquors and developing liquors, as may be most convenient and required for the work in hand. Such a machine is used in dyeing logwood black, aniline black, and many of the direct colours, etc. [Pg 80]

From the direct colours a large number of light shades are dyed on to cotton cloth by the process known as padding; this consists in passing the cloth through a liquor containing the dye-stuff, usually a little phosphate of soda is added, then between squeezing rollers, and finally drying the cloth. For this process there is used what is called a padding machine. This is shown in Figs. 25 and 26.

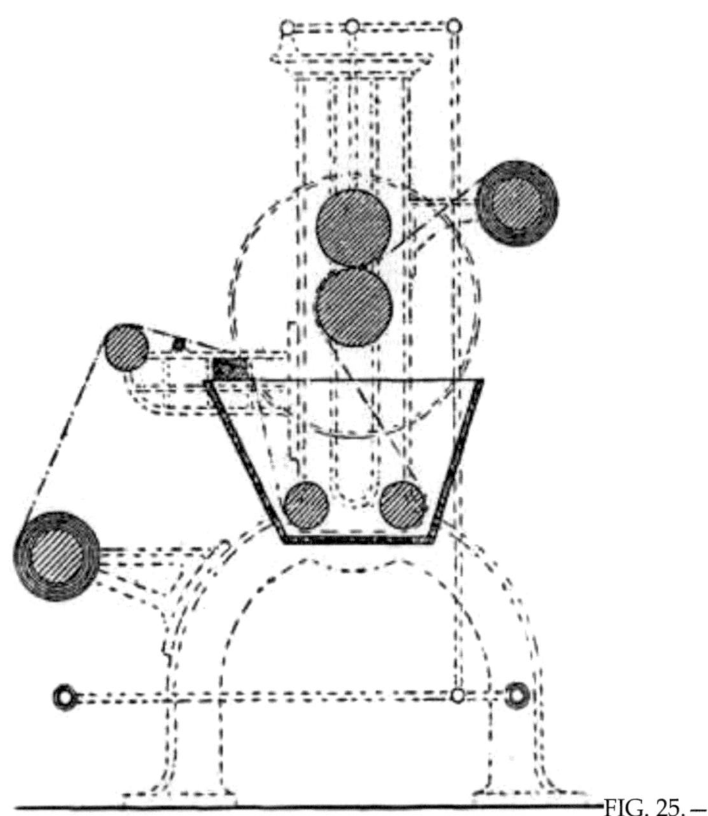 FIG. 25. —
Padding Machine.

It consists essentially of a trough, which contains two or more guide rollers, and in this is placed the padding liquor. Above the trough is fitted squeezing rollers, sometimes two as in Fig. 25, or three as in Fig. 26. Besides these, there [Pg 81]

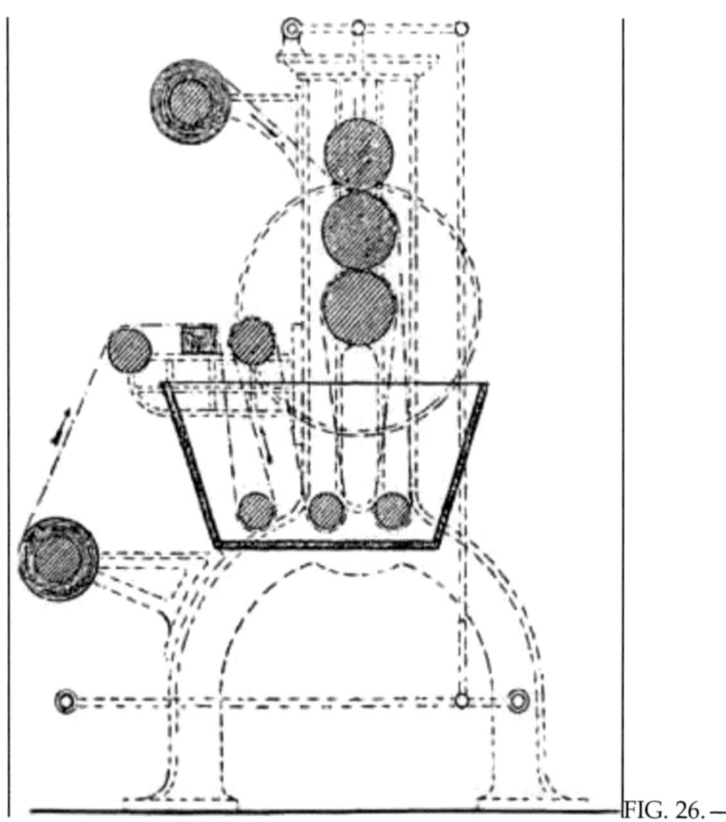

FIG. 26.—
Padding Machine.

are winding and beaming and other guide rollers. Fig. 25 shows the simplest padding machine, where the cloth passes once through the liquor and through the squeezing rollers. In Fig. 26 the cloth passes several times through the liquor and twice through the squeezing rollers, thus ensuring a more perfect impregnation of the cloth with the dye-liquor, and therefore a more uniform dyeing of the cloth. [Pg 82]

CHAPTER IV.

THE PRINCIPLES AND PRACTICE OF COTTON DYEING.

Students of cotton dyeing should have a good knowledge of the principles that underlie the processes of dyeing cotton fabrics. It is only by recognising these principles and then endeavouring to apply them to each individual case of dyeing, that the dyer or student will obtain a thorough grasp of his subject. It is the aim of the author to lay down these principles in a clear and intelligible form. Cotton is dyed in its loose raw condition, as yarn in the form of hanks, yarn in the form of cops, and in the woven pieces of every kind. Formerly the idea was prevalent among cotton dyers that the process which succeeds with piece goods would not answer with yarns. It is now recognised however that this is not so, that a process which will dye cotton yarn will also dye cotton piece goods or loose cotton. The differences which do exist in the practical working of the processes entirely arise from the difference in the form in which the cotton is presented to the dyer, for it must be obvious to any one that the mode of handling a piece of cotton cloth during the time it is in the dye-bath must be different from that of a hank of yarn, a parcel of loose cotton or a number of cops. The various machines used for dyeing all these forms and the manner of working them have been already described.

The dyes, whether natural — derived from the various dye-woods, etc. — or artificial — prepared from coal tar — may according to their varied chemical composition and consti [Pg 83] tution be divided into seventeen or eighteen distinct groups, but it is not intended here to give any account of them; the reader is referred to other books such as *The Dictionary of Coal Tar Colours*, by George H. Hurst; *The Chemistry of Coal Tar Colours*, by Benedikt and Knecht; or *The Chemistry of Organic Colouring Matters*, by Nietzki, where the composition and properties of the dyes are fully described.

From the manner in which the various dyes are applied to cotton, linen, wool and other fibres we can divide them into five groups, thus: —

Group   1.   Direct dyes.

"   2.   Basic dyes.

"   3.   Acid dyes.

"   4.   Mordant dyes.

          "   5.   Miscellaneous dyes.

First group, direct dyes, are now very numerous; they dye cotton, linen and other vegetable fibres from a plain bath, and do not require those fibres to be prepared in any way. Hence the reason of their being named direct, or by some the substantive colours. They will also dye wool and silk.

The second group, basic dyes, comprise some of the oldest of the coal-tar dyes; they dye wool and silk direct from plain baths, but require cotton, linen and other vegetable fibres to be previously prepared in baths of tannic acid, sumach or other tanning material.

The third group, acid dyes, are very numerous, and from both their chemical composition and mode of dyeing can be divided into several sub-groups. Their principal feature is that they dye wool and silk from baths containing Glauber's salt and some acid, hence their name of "acid dyes". They do not dye cotton or linen well, some not at all, others are absorbed to a slight extent by the cotton, but only pale tints [Pg 84] are produced, while others may be used along with metallic mordants to dye bright but pale and fugitive shades. The acid dyes comprise such as Acid green, Formyl violet, Acid magenta, Azo scarlet, Orange. Thiocarmine R., Patent blues, Wool greens, indigo extract, Eosines, etc.

The fourth group, mordant dyes, includes the alizarine series of coal-tar colours, logwood, Brazil wood and most natural colours, and some others. The principal feature of these dyes is that they require the cotton to be prepared with some metallic oxide, like those of chrome, alumina and iron, before dyeing, and the colour which is got depends partly upon the particular dye-stuff used and partly upon the oxide with which the cotton has been prepared.

The fifth group includes a few dyes like indigo, which are dyed on to cotton by various and special processes.

The processes of cotton dyeing employed to-day may be comprised under eight heads, namely: —

(1) Direct dyeing. (2) Direct dyeing followed by fixation with metallic salts. (3) Direct dyeing followed by fixation with developers. (4) Direct dyeing followed by fixation with couplers. (5) Dyeing on tannic mordant. (6) Dyeing on metallic mordant. (7) Developing the

colour direct on fibre. (8) Dyeing by impregnating the cotton with the dye-stuff, followed by oxidation or steaming.

It is of course not easy to elaborate a simple scheme of grouping the processes that shall definitely include all processes, but the above grouping will be found as successful as any, and each will be considered as fully as is deemed necessary. [Pg 85]

(1) DIRECT DYEING.

Nothing in the history of cotton dyeing caused such a revolution in the methods of working as did the introduction some fifteen years ago of the forerunners (Congo red, Benzo purpurine, Chrysamine, Azo blue) of the now numerous group of direct dyes, followed as they were by the Benzo, Congo, Diamine, Titan, Chicago and Zambesi members of the group. Prior to their introduction cotton dyeing was always more or less complicated and mordanting methods had to be employed.

With the introduction of the direct dyes cotton dyeing has become even more simple than wool or silk dyeing, and now all that is necessary is to prepare a dye liquor containing the necessary amount of dye-stuff and Glauber's salt, or common salt or soda, or some similar body, or a combination thereof. The method of working is to place the cotton in a lukewarm or even in a hot bath, raise to the boil, allow the goods to remain in the boiling bath for half an hour to an hour, then take them out, wring, wash and dry. This method is simple and will answer for all the dyes of this group. There are some that do not require the working to be done boiling; it is simply needful to enter the cotton into a boiling bath and work without steam until the bath has cooled down. Furious boiling is not needed—a gentle simmer gives the best results. Uneven dyeing seems to be an impossibility in this group of dyes, unless the cotton is dirty; no matter how the operations are carried on, level dyeing is the rule not the exception. An enormous variety of shades and tints can be obtained from these dyes, and they can be combined together in every conceivable manner and proportions.

No satisfactory explanation has yet been given as to what feature in the chemical composition of these dye-stuffs give them such an affinity for the cotton fibre as to enable [Pg 86] them to dye in so simple a manner such fast shades as they do; it is a fact there is such an affinity and there the matter must rest.

It has been found in practice that the efficiency of the dyeing operation depends, primarily of course, on the particular dye-stuff used, but also upon other factors, that a certain assistant be used. Some dyes work on the cotton better from a bath containing Glauber's salt, while with others common salt works best, while a little soda along with the salt facilitates the dyeing in some cases. It is practically impossible to specify here the best assistant for all the direct colours, on account of the great number of such dyes which are known, but in the practical recipes given below much useful advice will be found. Then the quantity of such assistants used is of much importance; there is one proportion at which the best results are obtained for each dye. The dyer should find out for himself by experiment and the use of the dyes he employs in his work what assistant and how much is best, and make his baths up to that strength. With some dyes 10 per cent. of the assistant will be found sufficient, while with others, 25 to 30 per cent. will not be too much. The percentage refers to the weight of the cotton that is taken.

One function of the assistants must be pointed out here: it is that in some cases they—especially the alkaline ones, soda, potash, borax, phosphate of soda—help the dyeing by promoting the solubility of the dye-stuff in the bath, thereby retarding the exhaustion of the bath and ensuring the production of level shades.

The following formulas show the application of the foregoing principles to the dyeing of numerous shades on to cotton and also the dyes which are applicable, some of the combinations which are possible with these direct dyes, and give some idea of the tints and shades of the colours that can [Pg 87] be got by their means. The best assistants to use are also indicated in the formulæ.

All the formulæ here given and all that will be given in future chapters are intended for 100 lb. weight of cotton fabrics in any condition, whether of loose cotton, yarn in cops, hanks or wraps and woven fabrics of every kind.

Bright Red.—Dye with 3 lb. Benzo purpurine 4 B, 3 lb. soda and 15 lb. Glauber's salt. This dye may also be used with 3 lb. soap and 10 lb. soda in the bath with equally good results.

Pale Salmon. — Prepare a dye-bath with 3 lb. salt, 5 lb. phosphate of soda, 1 lb. soap, ½ oz. Benzo orange R. For a pale shade like this it is not necessary to heat to the boil, a temperature of 170° to 180° F. is sufficient.

Dark Plum. — Prepare a dye-bath with 20 lb. of Glauber's salt, 2½ lb. soap, 1½ lb. Diamine black R O. and 2 lb. Diamine red N. Enter at 180° F., work for a few minutes, then raise to boil and dye to shade; lift, wash and dry.

Turkey Red. — Prepare a dye-bath with 1½ lb. Benzo purpurine 4 B, 1 lb. Brilliant purpurine, 2 lb. soap, 10 lb. borax. Enter the cotton at the boil and work for one hour; lift, wash and dry.

Lilac Red. — Prepare the dye-bath with 2 lb. soap, 5 lb. soda, 3 lb. Rose azurine G. Work at the boil for one hour.

Pink. — Prepare a bath containing 10 lb. soda, 1 oz. Rose azurine B. Enter at a boil and work for one hour, boiling to shade; lift, wash and dry.

Bordeaux. — Prepare a dye-bath with 15 lb. Glauber's salt, 5 lb. soda crystals, 3 lb. Diamine fast red F, 1 lb. Diamine violet N, 1 lb. Rose azurine G. Enter cold, then raise to the boil, and work for one and a half hours; lift, wash and dry.

Rose Pink. — The dye-bath is made with 2 lb. Erika B, 20 lb. Glauber's salt and 3 lb. soap, working at near the boil to shade. [Pg 88]

Brilliant Red. — Make the dye-bath with 24 lb. Brilliant purpurine R and 25 lb. Glauber's salt, working at the boil for one hour.

Deep Pink. — Make the dye-bath with ½ lb. Diamine rose B D, ½ lb. soda, 1 lb. soap and 5 lb. Glauber's salt, working at 150° F. for half an hour.

Dark Red. — Use in the dye-bath 3 lb. Diamine red 5 B, 2 lb. soda and 20 lb. Glauber's salt, working at the boil for one hour.

Pink. — Prepare the dye-bath with 4 oz. Diamine rose B D, 1 lb. Turkey-red oil, 40 lb. Glauber's salt. Dye at the boil for one hour.

Scarlet. — Prepare the dye-bath with 4 lb. Diamine scarlet 3 B, 1 lb. Turkey-red oil, 20 lb. Glauber's salt. Dye at the boil for one hour.

Scarlet.—Prepare the dye-bath with 3 lb. Titan scarlet C, ½ lb. Titan orange, 50 lb. salt. Work at the boil for thirty minutes, then lift, wash and dry. The dye-bath is not exhausted and may be used for further lots.

Crimson Red.—Prepare the dye-bath with 5 lb. Titan scarlet D and 50 lb. salt. Work at the boil for fifty minutes, then lift, wash and dry. The bath is not exhausted, the cotton taking up only about 3 lb. of the dye-stuff; it may therefore be kept for further use, when for each succeeding lot 3 to 3½ lb. of colour and 25 lb. of salt only need be added.

Scarlet.—Prepare the dye-bath with 5 lb. Rosophenine 5 B, dissolved in 50 gallons hot water, 2 gallons caustic soda lye (60° Tw.). When thoroughly dissolved add 150 lb. salt. Make up the bath to 100 gallons. Enter the yarn and work for a quarter to half an hour at about 180° F; squeeze off and wash thoroughly in cold water until the water runs off clean.

Rose Red.—Use 1 lb. Diamine red 10 B, 3 lb. soda, and 20 lb. Glauber's salt. [Pg 89]

Deep Crimson.—Use 3 lb. Diamine red 10 B, 3 lb. soda and 20 lb. Glauber's salt.

Claret.—Use 3 lb. Diamine Bordeaux B, 3 lb. soda and 20 lb. Glauber's salt.

Pink.—The dye-bath is made with 5 oz. Dianil red 4 B, 5 lb. salt and 3 lb. soda.

Scarlet.—Use in the dye-bath 3 lb. Dianil red 4 B, 15 lb. salt and 5 lb. soda. Work at the boil for one hour.

Claret.—Dye with 1½ lb. Dianil claret G, 3 lb. soda and 20 lb. salt. Work at the boil for one hour.

Maroon.—Dye with 3 lb. Dianil claret B, 3 lb. soda and 20 lb. salt. Work at the boil for one hour.

Bright Scarlet.—Use in the dye-bath 2½ lb. Dianil red 4 B 5 oz. Dianil orange G, 3 lb. soda and 15 lb. salt.

Dark Maroon.—Make the dye-bath with 1 lb. Dianil red 4 B, 2 lb. Dianil claret G, 13 oz. Dianil claret B, 5 lb. soda and 20 lb. salt.

Crimson. — Dye with 3 lb. Congo rubine, 5 lb. soda and 20 lb. Glauber's salt.

Dark Maroon. — Use in the dye-bath 1 lb. Benzo purpurine 4 B, 3 lb. Congo Corinth G, 3 lb. soda and 20 lb. Glauber's salt, working at the boil to shade.

Pale Fawn Red. — Use in the dye-bath 1½ oz. Diamine red 5 B, 1½ oz. Diamine catechine G, 3 lb. soda and 10 lb. Glauber's salt.

Rose Red. — Prepare the dye-bath with ¾ lb. Diamine Bordeaux B, 3 oz. Diamine orange B, 3 lb. soda and 20 lb. salt.

Crimson. — Use in the dye-bath ¾ lb. Diamine Bordeaux B, 3 oz. Diamine fast yellow B, 3 lb. soda and 20 lb. Glauber's salt.

Salmon. — Dye with 1½ oz. Diamine fast red F, 1½ oz. Diamine fast yellow B, 3 lb. soda and 10 lb. Glauber's salt. [Pg 90]

Terra-Gotta Red. — Dye with 1½ lb. Diamine brown M, ¾ lb. Diamine fast red F, 3 lb. soda and 20 lb. Glauber's salt.

Lilac Red. — Dye with 4 lb. Heliotrope B B, 3 lb. soda and 15 lb. Glauber's salt.

Bright Pink. — Use in the dye-bath 2 oz. Rose azurine G, 1 lb. soda and 10 lb. Glauber's salt. Nearly all the direct reds give good pink tints when used in proportion, varying from 0.1 to 0.25 per cent. of dye-stuff.

Bright Straw. — Dye in a bath made of ¼ lb. Titan yellow G G, 10 lb. salt, for three-quarters of an hour, then lift, wash and dry.

Yellow. — Prepare a dye-bath with 1 lb. Titan yellow Y, 10 lb. salt. Heat to 180° F., enter the goods, raise to boil, and dye for one hour; lift, wash and dry.

Yellow. — Prepare the dye-bath with ¼ lb. Diamine fast yellow A, 1 lb. Turkey red oil, 20 lb. Glauber's salt. Dye at the boil for one hour.

Sun Yellow. — Prepare the dye-bath with 2 lb. Sun yellow, 30 lb. common salt. Dye at the boil. The bath is kept for further lots.

Yellow. — Prepare the dye-bath with 1 lb. Direct yellow R, 20 lb. Glauber's salt. Dye at the boil for one hour.

Yellow.—Prepare the dye-bath with 2 lb. Curcuphenine, 20 lb. common salt. Work at the boil for one hour; lift, rinse and dry.

Old Gold.—Make the dye-bath with 5 lb. Diamine yellow N powder, 20 lb. phosphate of soda, 10 lb. soap. Work at the boil for one hour; finish in the usual way. The bath may be kept for other lots of goods.

Dark Yellow.—The bath is made from 2 lb. Toluylene orange G, 10 lb. phosphate of soda, and 2½ lb. soap, working at the boil to shade.

Bright Yellow.—Use 1 lb. Chrysophenine, 2 lb. phosphate of soda and 10 lb. Glauber's salt. [Pg 91]

Lemon Yellow.—Use 1 oz. Chrysamine G, 2 lb. phosphate of soda and 10 lb. Glauber's salt.

Yellow.—Dye with 2 lb. Oxyphenine and 20 lb. salt.

Yellow Olive.—Use in the dye-bath 2 oz. Cotton brown N, 4½ oz. Diamine bronze G, 4½ oz. Diamine fast yellow B, 3 lb. soda and 20 lb. salt.

Green Yellow.—Dye with ½ lb. Diamine fast yellow B. 2 oz. Diamine bronze G, 3 lb. soda and 10 lb. Glauber's salt.

Gold Yellow.—Use in the dye-bath 3 lb. Columbia yellow, 3 lb. soda and 20 lb. Glauber's salt.

Cream.—Dye with ½ oz. Toluylene orange G, 24 grains Brilliant orange G, 1 lb. soda and 10 lb. Glauber's salt.

Primrose.—Dye with 3 oz. Dianil yellow 3 G, 2 lb. soda and 10 lb. salt.

Gold Yellow.—Dye with 2½ lb. Dianil yellow G, ½ lb. soda and 15 lb. salt.

Buff Yellow.—Dye with 3½ oz. Dianil yellow 2 R, ½ lb. soda and 10 lb. salt.

Orange.—Prepare the dye-bath with 2 lb. Chlorophenine orange R, 20 lb. common salt. Work at the boil for one hour; lift, rinse and dry.

Red Orange.—Make the dye-bath with 3 lb. Mikado orange 4 R and 25 lb. salt. Work at the boil for one hour.

Orange.—Make the dye-bath with 3 lb. Mikado orange G and 25 lb. salt. Work at the boil for one hour.

Pale Orange.—The dye-bath contains 6 oz. Diamine Orange G, 1½ oz. Diamine fast yellow B, ¼ oz. Diamine scarlet B, 3 lb. soda and 15 lb. Glauber's salt.

Olive Yellow.—Dye with ¾ lb. Diamine fast yellow B, 1 oz. Oxydiamine black N, 1½ oz. Diamine bronze G, 3 lb. soda and 20 lb. Glauber's salt.

Dark Orange.—Dye with 3 lb. Columbia orange R, 3 lb. soda and 20 lb. Glauber's salt at the boil for one hour. [Pg 92]

Bright Orange.—Use 3 lb. Congo orange R, 3 lb. soda and 20 lb. Glauber's salt at the boil for one hour.

Pale Orange.—Dye with 3 lb. Dianil orange 2 R, 2 lb. soda and 10 lb. salt at the boil for one hour.

Brilliant Orange.—Dye with 4 lb. Dianil orange G and 20 lb. salt for one hour.

Deep Orange.—Dye with 2 lb. Oxydiamine orange R, ¾ lb. soda and 20 lb. salt for an hour.

Pale Orange.—Dye with ¾ lb. Diamine fast yellow B, 1 lb. Diamine orange B, 3 lb. soda and 15 lb. Glauber's salt.

Bright Orange.—Dye with 1½ lb. Benzo orange R, 1½ lb. Chrysamine R, 10 lb. phosphate of soda and 2 lb. soap.

Green.—Prepare the dye-bath with 2 lb. Benzo green G, 10 lb. Glauber's salt. Enter lukewarm, bring slowly to the boil, dye for one hour at the boil.

Russian Green.—Make the dye-bath with 16 oz. Diamine black H W, 4 oz. Diamine fast yellow A, 3 lb. soda, 15 lb. Glauber's salt, working at the boil for one hour, then lift, wash and dry.

Dark Olive.—Prepare a dye-bath with 3½ lb. Benzo olive, 2½ lb. Diamine black B O, 2 lb. Diamine yellow, 20 lb. common salt, 2 lb. soap. The goods are entered into the bath at 160° F., then heat is raised to the boil, and the dyeing continued for one hour, then lift, wash and dry.

Dark Olive.—Dye in a bath of 2 lb. Titan yellow Y, 1 lb. Diamine brown Y, 1½ lb. Diamine blue 3 B, 2 lb. soda. Work for one hour, then lift, wash and dry.

Olive.—Prepare a dye-bath with 15 lb. phosphate of soda, 3 lb. soap, 1½ lb. Diamine yellow N, 4 oz. Diamine blue 3 B, 1½ oz. Diamine brown V. Dye at the boil to shade; lift, wash and dry.

Green Olive.—Prepare the dye-bath with 1 lb. Diamine black R O, 1 lb. Chrysamine, ¼ lb. Benzo brown, 5 lb. soda, 5 lb. salt, 2 lb. soap. The goods are entered at about 180° [Pg 93] F. and worked for a short time, then the temperature is raised to the boil, and the goods are worked for one hour, lifted, washed and dried.

Reseda.—Prepare a bath with 10 lb. Glauber's salt, 2 lb. soap, ½ lb. Diamine black R O, 2 lb. Diamine yellow N. Enter at 120° F., heat to boil and dye for one hour at that temperature; lift, wash and dry.

Sage Green.—Prepare a dye-bath with 10 lb. Glauber's salt, ½ lb. Diamine black R O, 2 lb. Diamine yellow N. Enter at about 150° F. and then raise to boil and dye boiling for one hour, wash and dry.

Drab.—Prepare the dye-bath with 10 lb. Cross dye drab, 5 lb. soda crystals. Enter at the boil and work at this temperature for half an hour. Whilst dyeing add gradually 75 lb. salt. Rinse well and dry.

Olive.—Prepare the dye-bath with 2 lb. Dianil olive, 5 lb. phosphate of soda, 5 lb. common salt. Dye at the boil for one hour.

Olive.—The dyeing is done in a bath containing 4 oz. Diamine black H W, 1¾ lb. Diamine bronze G, 5 lb. soda, 15 lb. Glauber's salt. Work at the boil for one hour.

Grass Green.—Make the dye-bath with 2 lb. Chrysamine G, 1½ oz. Benzo azurine G, 3 lb. soap and 10 lb. borax, working at the boil for one hour.

Green.—Make the dye-bath with 2 lb. Titan yellow Y, 1 lb. Titan blue 3 B and 20 lb. salt.

Bright Grass Green.—Dye for an hour at the boil with 1 lb. Sulphon azurine D, 2 lb. Thiazole yellow and 20 lb. Glauber's salt.

Green.—Use in the dye-bath 3 lb. Diamine green B, 3 lb. soda and 20 lb. Glauber's salt, working at the boil to shade.

Dark Green.—Dye with 3 lb. Diamine dark green N, 3 lb. soda and 20 lb. Glauber's salt. [Pg 94]

Green.—Use in the bath 3 lb. Benzo green B B, 3 lb. soda and 20 lb. Glauber's salt at the boil for one hour.

Dark Sea Green.—Dye with 5 oz. Diamine black H W, 3 oz. Diamine catechine G, 3 oz. Diamine fast yellow B, 3 lb. soda and 10 lb. Glauber's salt.

Pale Green.—Use in the dye-bath 3 lb. Diamine fast yellow B, 2 oz. Diamine black H W, 3 lb. soda and 10 lb. Glauber's salt.

Bright Pea Green.—Use in the dye-bath 1 oz. Thioflavine S, ¼ oz. Diamine sky blue F F and 20 lb. Glauber's salt.

Dark Green.—Use 1¾ lb. Diamine green G, ¾ lb. Oxydiamine yellow G G, 3 lb. soda and 20 lb. Glauber's salt, working at the boil for one hour.

Deep Green.—Use 1¾ lb. Diamine green G, ¾ lb. Diamine black B H, ½ lb. Oxydiamine yellow G G, 3 lb. soda, and 20 lb. Glauber's salt.

Sea Green.—Use 2 oz. Dianil yellow R, 2½ oz. Dianil blue B, 1¾ oz. Dianil dark blue R, 1 lb. soda, and 20 lb. salt, working at the boil.

Leaf Green.—Dye with 1½ lb. Dianil yellow 3 G, 1 lb. Dianil blue B, 11 oz. Dianil blue 2 R, 3 lb. soda, and 20 lb. salt at the boil for one hour.

Deep Green.—Dye with 2½ lb. Dianil yellow 3 G, 2½ lb. Dianil blue 2 R, 6 oz. Dianil dark blue R, 3 lb. soda, and 20 lb. salt at the boil for one hour.

Greens are largely produced by mixing yellows and blues together as will be seen from the recipes given above; the particular shade of green which is got from a combination of blue and green depends upon the quality of the dye-stuffs used: thus, to produce bright greens of a pure tone, it is essential that the yellow used shall have a greenish tone like Thioflavine S, Thiazole yellow, or Dianil yellow 3 G, while the blue must also have a greenish tone like Diamine [Pg 95] sky blue, Benzo blue 3 B, etc. By using yellows like Diamine fast yellow R, and dark blues like Benzo azurine 3 R, Diamine blue R W, Dianil dark blue R, the green which is got is darker and duller in tone. The addition of such a dye as Diamine black B H throws the shade more on to an olive, while a brown dye-stuff, like Diamine brown M, or an orange dye, like Titan orange N, throws the green on to a sage tone. Examples of these effects will be found among the recipes given above.

It may be added here that by using smaller quantities, but in the same proportions as given in the above recipes, a great range of tints and shades of green can be dyed from very pale to very deep.

Bright Blue.—Prepare a dye-bath with ½ lb. Congo blue 2 B, 5 lb. salt, 5 lb. phosphate of soda, 2 lb. soap. Work at the boil for one hour, then rinse and dry.

Dark Navy.—Prepare a dye-bath with 1 lb. Diamine black R O, 2 lb. Diamine blue 3 R, 8 lb. Glauber's salt, 2 lb. soap. Enter the cotton at 180° F., and boil for one hour.

Pale Blue.—Prepare a dye-bath with 10 lb. salt, 3 lb. soda, 3 oz. diamine blue 3 R. Work for one hour at the boil, then lift, wash and dry.

Sky Blue.—Prepare a dye-bath with 2 lb. Titan como G, 20 lb. common salt, 2 oz. acetic acid. Work at the boil for half an hour, then lift, wash and dry.

Bright Blue.—Prepare the dye-bath with 1½ lb. Chicago blue 6 B, 20 lb. Glauber's salt, 3 lb. soap. Work at the boil for one hour, then lift, wash and dry.

Pale Sky Blue.—Make the dye-bath with 1 oz. Chicago blue 6 B, 10 lb. Glauber's salt, 2 lb. soap. Work at the boil for one hour, then lift, wash and dry.

Sky Blue.—Prepare the dye-bath with 1 lb. Diamine sky blue F F, 1 lb. Turkey-red oil, 20 lb. Glauber's salt. Dye at the boil for one hour.

Dark Blue.—Prepare the dye-bath with 2½ lb. Diamineral [Pg 96] blue R, 2½ lb. Diamine deep black Cr, 1 lb. Turkey-red oil, 40 lb. Glauber's salt. Dye at the boil for one hour.

Dark Blue.—Prepare the dye-bath with 3 lb. Triamine black B, 15 lb. Glauber's salt, in 50 gallons of water. Enter at 150° F., and boil for one hour. Allow the goods to remain until the water is cold, when the dye-bath will be completely exhausted.

Blue.—Prepare the dye-bath with 2 lb. Diamine steel blue L, 2 lb. soda, 15 lb. Glauber's salt. Dye at the boil for one hour.

Blue.—Prepare the dye-bath with 4 lb. Diamine blue B G, 2 lb. soda, 20 lb. Glauber's salt. Dye at the boil for one hour. In shade this is very similar to that got with Diamine brilliant blue G, which however should be used for light shades on account of its brightness. For deep shades Diamine blue B G, is preferable, because of its greater tinctorial power.

Light Indigo Blue.—Prepare the dye-bath with 1 lb. Paramine indigo blue, 2 lb. soda, 20 lb. Glauber's salt. Enter at about 150° F., and dye at the boil for one hour.

Navy Blue.—Prepare the dye-bath with 4 lb. Paramine navy blue R, 2 lb. soda, 20 lb. Glauber's salt. Enter at about 150° F., and dye at the boil for one hour.

Blue.—Prepare the dye-bath with 1 lb. Paramine navy blue R, 2 lb. soda, 20 lb. Glauber's salt. Enter at about 150° F., and dye at the boil for one hour.

Navy Blue.—Prepare the dye-bath with 4 lb. Benzo chrome black blue B, 15 lb. Glauber's salt, 3 lb. soda. Work at the boil for one hour; lift, rinse and dry.

Grey Blue.—Prepare the dye-bath with 2 lb. Paramine blue black S, 2 lb. soda, 20 lb. Glauber's salt. Enter at 150° F., and dye for one hour at boil.

Blue.—Prepare the dye-bath with 1 lb. Paramine blue B, 2 lb. soda, 20 lb. Glauber's salt. Enter at about 150° F., and dye at the boil for one hour. [Pg 97]

Slate Blue.—Prepare the dye-bath with ¼ lb. Diamine black B H, ¾ oz. Diamine fast yellow B, 2 lb. soda, and 10 lb. Glauber's salt. Dye at the boil to shade.

Deep Blue.—Use 3¼ lb. Diamine blue B X, ½ lb. Oxydiamine black N, 3 lb. soda and 20 lb. Glauber's salt at the boil for one hour.

Blue.—Dye at the boil for one hour with 1½ lb. Diamine sky blue, 2 oz. Diamine green B, 2 lb. soda and 10 lb. Glauber's salt.

Navy. — Dye with 1 lb. Dianil dark blue R, 8 oz. Dianil black C R, 5 lb. soda and 20 lb. salt at the boil for one hour.

Dark Navy. — Use 2 lb. Dianil blue B, 2 lb. Dianil dark blue R, ¾ lb. Dianil black C R, 2 lb. soda and 25 lb. salt, working at the boil for one hour.

Deep Blue. — Dye with 3½ lb. Diamine blue black E, 5 lb. soda and 20 lb. Glauber's salt at the boil for one hour.

Deep Blue. — Dye with 3 lb. Zambesi black B R, 3 lb. soda and 20 lb. Glauber's salt at the boil for one hour.

Dark Navy. — Use 3 lb. Dianil dark blue R, 3 lb. caustic soda 70° Tw., and 25 lb. salt, working at the boil for one hour.

Violet Blue. — Dye with 3 lb. Dianil dark blue 3 R and 25 lb. salt at the boil for one hour.

Bright Blue. — Use 1 lb. Dianil blue B, and 20 lb. salt, working at the boil for one hour.

Full Blue. — Dye with 3 lb. Brilliant azurine 5 G, 5 lb. common salt, 5 lb. phosphate of soda and 2 lb. soap at the boil for one hour.

Dark Blue. — Dye with 3 lb. Erie blue B X, 3 lb. soda and 20 lb. Glauber's salt at the boil for one hour.

Pale Blue. — Dye with 1 lb. Chicago blue 6 B, 3 lb. soda and 20 lb. salt at the boil for one hour.

Deep Blue. — Dye with 1½ lb. Oxydiamine black A, 2 lb. Diamine deep blue R, 3 lb. soda and 20 lb. Glauber's salt at the boil for one hour. [Pg 98]

Blue. — Dye with 6 oz. Diamine blue 3 B, 1½ oz. Diamine sky blue F F, 3 lb. soda and 10 lb. Glauber's salt.

Navy. — Dye with 4 lb. Diamine new blue R, 3 lb. soda and 20 lb. Glauber's salt at the boil for one hour.

Dark Navy. — Dye with 1¾ lb. Diamineral blue R, 3 lb. soda and 20 lb. Glauber's salt at the boil for one hour.

Sky Blue. — Prepare the dye-bath with 6 oz. Diamine sky blue F F, 3 lb. soda and 10 lb. Glauber's salt, working at the boil for one hour.

Dark Blue.—Use in the dye-bath 3 lb. Diamine blue R W, 2 lb. soda and 20 lb. Glauber's salt, working at the boil for one hour.

Dark Blue.—Prepare the dye-bath with 3 lb. Triamine black B T, and 15 lb. Glauber's salt; work at the boil to shade.

Blue.—Use 2 lb. Direct indigo blue and 15 lb. Glauber's salt; work at the boil.

Bright Blue.—Use in the dye-bath 3 lb. Titan como S N, 2 lb. acetic acid and 20 lb. salt, working at the boil for one hour.

Turquoise Blue.—Dye with 1 lb. Diamine sky blue, 1 oz. Diamine fast yellow B, 2 lb. soda and 10 lb. Glauber's salt, working at the boil to shade.

Dark Navy.—Use 4 lb. Titan navy R, and 20 lb. salt at the boil for one hour.

Green Blue.—Dye with 1 lb. Dianil blue G, 2½ oz. Dianil yellow G, 1 lb. soda and 20 lb. salt at the boil for one hour.

Many more formulæ could have been given, but the above will perhaps suffice; they include all the best of the direct blues. Paler tints of blue may be got by using from 1 to 2 per cent. of any of these blues and also of the many direct blacks now on the market. The direct blues as a rule have a good degree of fastness to light.

Lilac.—Prepare a dye-bath with ¼ lb. Hessian brown 2 M, [Pg 99] 1 oz. Azo mauve A M, 1 lb. soap, 2 lb. soda, 10 lb. salt. Work at the boil for one hour, then lift, wash and dry.

Plum.—Dye with 3 lb. Oxydiamine violet G, 3 lb. soda and 20 lb. salt.

Dark Plum.—Use in the dye-bath 3 lb. Oxydiamine violet B, 3 lb. soda and 20 lb. Glauber's salt, working at the boil.

Violet.—Make the dye-bath with ¾ lb. Oxydiamine violet B, 1 lb. soda and 10 lb. Glauber's salt, and dye at the boil to shade.

Violet.—Dye with 12 oz. Dianil blue 4 R, 2 oz. Dianil blue B and 10 lb. salt at the boil.

Lilac.—Dye with 1½ oz. Diamine rose G D, ¾ oz. Diamine sky blue F F, 1 lb. soda and 10 lb. Glauber's salt at the boil to shade.

Red Violet. — Make the dye-bath with ½ lb. Diamine violet N, ½ oz. Diamine brilliant blue G, 1 lb. soda and 10 lb. Glauber's salt, working at the boil.

Red Violet. — Dye with 1 lb. Diamine violet N, 1 lb. soda and 10 lb. Glauber's salt.

Bright Red Lilac. — Dye with 1¾ lb. Erika B N, 4 oz. Chicago blue 4 R, 3 lb. soda and 20 lb. Glauber's salt at the boil.

Grey Lilac. — Dye with 12 oz. Neutral grey G, 3 oz. erika B N, 1 lb. soda and 10 lb. Glauber's salt at the boil for one hour.

Pale Lilac. — Dye with 2½ oz. Dianil claret B, 2½ oz. Dianil blue 4 R, and 10 lb. salt.

Light Plum. — Dye with 10 oz. Dianil claret B, 10 oz. Dianil blue and 20 lb. salt.

Dull Lilac. — Dye with ½ lb. Diamine brown V, 1 lb. soda and 10 lb. Glauber's salt.

Heliotrope. — Dye with 4 oz. Heliotrope 2 B, 1 lb. soda and 10 lb. Glauber's salt.

Plum. — Dye with 3 lb. Congo Corinth B, 3 lb. soda and 10 lb. Glauber's salt. [Pg 100]

Dull Violet. — Use in the dye-bath 1½ lb. Chicago blue 4 R, 14 oz. Erika B N, 3 lb. soda and 20 lb. Glauber's salt, working at the boil for one hour.

Red Lilac. — Dye with 6 oz. Oxydiamine violet G, 2 oz. Oxydiamine violet B, 1 lb. soda and 10 lb. Glauber's salt at the boil for one hour.

Violet. — Dye with 3 oz. Diamine violet N, 2 oz. diamine blue 3 R, 1 lb. soda and 10 lb. Glauber's salt.

Fawn Drab. — Prepare a dye-bath with 1 lb. Cachou de laval, ¼ oz. Benzo purpurine B. Enter the cotton into this bath in the cold and heat to the boil, taking about one hour for the operation, then add 4 lb. common salt and boil for three-quarters of an hour longer; lift, wash and dry.

Pale Olive Brown. — The dye-bath is made with 1 lb. Diamine bronze G, 1 oz. Cotton brown N, 3 oz. Diamine gold, 5 lb. soda, 15 lb. Glauber's salt. Work at the boil for one hour, then lift, wash and dry.

Red Brown.—Prepare a dye-bath with 1¾ lb. Cotton yellow, 4 lb. Hessian brown 2 B N, 2 lb. Diamine black R O, 1 lb. soda, 2 lb. salt. Enter the goods at 180° F., then raise to the boil and work to the shade; lift, wash and dry.

Brown Drab.—Prepare a dye-bath with ¼ lb. Cotton brown N, ¾ oz. Diamine yellow N, ¾ oz. Diamine black B O, 15 lb. phosphate of soda, 3 lb. soap. Work at the boil for one hour.

Gold Brown.—Prepare the dye-bath with 16¼ oz. Toluylene orange G, 9½ oz. Toluylene orange R, 4¾ oz. azo mauve, 2½ lb. soap, 5 lb. soda. Dye at the boil for one hour.

Chestnut Brown.—Prepare a dye-bath with 10 lb. common salt, 2 lb. Benzo brown G, ½ lb. Benzo azurine G, ½ lb. Chrysophenine. Enter the goods at 150° F., raise to the boil and dye boiling for one hour.

Purple Brown.—Prepare a dye-bath with 10 lb. common salt, 2 lb. Benzo brown N B, 1 lb. Azo violet. Enter the [Pg 101] cotton at 150° F., raise to the boil and dye boiling for an hour; lift, wash and dry.

Brown.—Prepare a dye-bath with 5 lb. soda, 10 lb. Glauber's salt, 12 oz. Chrysamine, 1 oz. Benzo purpurine, 6¼ oz. Benzo azurine. Dye at the boil for one hour, rinse and dry. The brown thus got is fast to washing.

Dark Chestnut Brown.—Prepare a dye-bath with 10 lb. salt, 3 lb. Benzo brown N B X, raise to 150° F., enter goods, heat to boil, and work for one hour; lift, rinse and dry.

Dark Brown.—Prepare a dye-bath with 20 oz. Glauber's salt per gallon of water used, 2½ lb. soap, 1½ lb. Diamine black R O, 2 lb. Cotton brown N. Enter the yarn at 180° F., give three turns, raise temperature to boil, and work to shade; lift, rinse and wash.

Gold Brown.—Prepare the dye-bath with 4 lb. Titan gold, 50 lb. salt. Work at the boil for thirty minutes, then lift, wash and dry. The dye-bath is not exhausted, only about 3 lb. of the colour being taken up by the cotton. It may, therefore, be kept for further lots, adding 3 lb. more colour and about 20 lb. more salt for each batch of cotton, or if it is not desired to keep the bath, add less colour to start with, and towards the end of the operation add more salt.

Brown. — Prepare the dye-bath with 4 lb. Paramine brown G, 20 lb. Glauber's salt, 2 lb. soda. Dye at the boil for one hour.

Light Brown. — Prepare the dye-bath with 3 lb. Diamine catechine G, 3 lb. soda, 15 lb. Glauber's salt. Dye at the boil for one hour.

Dark Brown. — Prepare the dye-bath with 5 lb. Diamine catechine B, 3 lb. soda, 15 lb. Glauber's salt. Dye at the boil for one hour.

Dark Drab. — Prepare the dye-bath with 1 lb. Titan brown Y, 3 oz. Columbia green, 32¾ oz. Diamine bronze, 17 lb. Glauber's salt. Work at the boil for one hour, then lift, wash and dry. [Pg 102]

Pale Brown. — The dye-bath is made with 2 lb. Mikado orange 4 R, 3 oz. Benzo fast grey, 30 lb. Glauber's salt. Work at the boil for one hour, then lift, ash and dry.

Gold Brown. — Make a dye-bath with 1 lb. Titan gold, 50 lb. common salt. Enter at the boil, work for an hour, then lift, wash and dry. Keep the bath for another lot of goods; it will only require the addition of about 14 oz. of colour and 10 lb. salt.

Buff Brown. — Make the dye-bath with ¾ lb. Titan gold, ¼ lb. Titan brown R, 5 oz. Titan blue 3 B, 40 lb. common salt. Work at the boil to shade, then lift, wash and dry.

Deep Chestnut Brown. — Make the dye-bath with 3 lb. Titan brown R, 1½ lb. Titan blue R, 25 lb. common salt. Work at the boil for an hour, then lift, wash and dry.

Light Seal Brown. — Make the dye-bath with 10 lb. salt, 2 lb. soda, 14 oz. Oxyphenine, ¾ lb. Atlas red R, 6 oz. Diamine blue B X. Work at the boil to shade, then lift, wash and dry.

Orange Brown. — Make a dye-bath with 10 lb. salt, 2 lb soda, 14 oz. Oxyphenine, 1 lb. Atlas red R, 1 oz. Diamine blue B X. Work at the boil to shade, then lift, wash and dry.

Pale Nut Brown. — Use in the dye-bath 4½ oz. Diamine catechine G, 1 oz. Diamine brown M, 1 oz. Diamine catechine B, 2 lb. soda and 10 lb. Glauber's salt, working at the boil for one hour.

Walnut Brown.—Dye with 1 lb. Diamine brown M, 3 oz. Diamine orange G, 2 oz. Diamine black H W, 2 lb. soda and 10 lb. Glauber's salt at the boil for one hour.

Black Brown.—Use in the dye-bath 3 lb. Diamine brown M, ¾ lb. Diamine blue black R, 3 lb. soda and 20 lb. Glauber's salt, working at the boil.

Reddish Brown.—Dye with 2 lb. Dianil brown R, 5 lb. salt and 5 lb. phosphate of soda at the boil for one hour.

Chocolate Brown.—Dye with 2 lb. Dianil brown T, 5 lb. phosphate of soda and 5 lb. salt at the boil for one hour. [Pg 103]

Dark Brown.—Dye with 2 lb. Dianil dark brown, 5 lb. salt and 5 lb. phosphate of soda at the boil for one hour.

Light Brown.—Prepare the dye-bath with 5 lb. Diamine catechine G, 3 lb. soda and 15 lb. Glauber's salt.

Brown.—Dye with 2¼ lb. Cotton brown N, 4 oz. Diamine black H W, 2 lb. soda and 20 lb. Glauber's salt at the boil for one hour.

Dark Walnut Brown.—Make the dye-bath with 3¼ lb. Diamine brown M, 6 oz. Diamine catechine B, 6 oz. Diamine red 5 B, 2 lb. soda and 20 lb. Glauber's salt; work at the boil.

Dark Chestnut Brown.—Dye with 2½ lb. Dianil brown R, 1 lb. soda and 20 lb. salt at the boil.

Dark Brown.—Dye with 2 lb. Dianil brown 3 G O, 2 lb. Dianil brown B D, 1 lb. Dianil red 4 B, 3 lb. soda and 25 lb. salt at the boil for one hour.

Brown.—Prepare the dye-bath with 5 lb. Mikado brown M and 25 lb. salt; work the cotton in this at the boil for one hour.

Nut Brown.—Use in the dye-bath 2½ lb. Benzo brown G and 15 lb. salt, working at the boil.

Dark Brown.—Use in the dye-bath 3 lb. Benzo brown N B and 15 lb. Glauber's salt, working at the boil.

Dark Brown.—Make the dye-bath with 4 lb. Diphenyl brown B N, 10 lb. Glauber's salt and 4 lb. soap, working at the boil to shade.

Black Brown.—Use in the dye-bath 2½ lb. Dianil brown 3 G O, 1½ lb. Dianil brown G, ¾ lb. Dianil dark blue R, 3 lb. soda and 25 lb. salt.

Dark Brown.—Dye with 1½ lb. Zambesi black D, 1¼ lb. Brilliant orange G, 3 lb. soda and 20 lb. Glauber's salt at the boil to shade.

Gold Brown.—Dye with 2 lb. Curcumine S, 1 lb. Columbia orange R, 5 oz. Columbia black F B, 3 lb. soda and 15 lb. Glauber's salt at the boil. [Pg 104]

Dark Chestnut.—Dye at the boil with 2 lb. Columbia Orange R, 8 oz. Columbia black F B, 2 lb. soda and 10 lb. Glauber's salt.

Sage Brown.—Dye with 1 lb. Zambesi black D, 1 lb. Curcumine S, ¾ lb. Diamine orange G D, 3 lb. soda and 30 lb. Glauber's salt at the boil.

Deep Brown.—Dye 3½ lb. Diamine brown M, ¾ lb. Oxydiamine orange G, ¾ lb. Diamine black H W, 3 lb. soda and 20 lb. Glauber's salt at the boil.

Chestnut.—Dye with 2½ lb. Diamine brown G, ¾ lb. Oxydiamine orange R, 3 lb. soda and 20 lb. Glauber's salt.

Pale Walnut Brown.—Dye with 3½ lb. Diamine brown M, ¾ lb. Oxydiamine orange G, 1½ oz. Diamine black B H, 3 lb. soda and 20 lb. Glauber's salt.

Various other browns may be obtained by combining the various direct browns together or with other direct dyes. The use of a yellow or orange will brighten them; that of a red will redden the shade; the addition of a dark blue or a black will darken the shade considerably. It may be useful to remember that a combination of red, orange and blue or black produces a brown, and by using various proportions a great range of shades can be dyed.

Black.—Prepare a dye-bath with 6 lb. Diamine black R O, 2 oz. Thioflavine S, 2 lb. soap, 10 lb. salt. Enter the cotton at the boil and dye for one hour; lift, wash and dry.

Black.—Prepare the dye-bath with 5 lb. Direct deep black E extra, and ½ to 1½ oz. common salt per gallon of water. Dye at the boil for one hour.

Deep Black.—Prepare the dye-bath with 5½ lb. Diamine deep black R B, 2 lb. soda, 20 lb. Glauber's salt. Dye at the boil for one hour; lift, wash and dry.

Black.—Prepare the dye-bath with 5 lb. Direct triamine black G X, 15 lb. Glauber's salt. Dye for one hour at the [Pg 105] boil; lift, rinse and dry. In working for from two to three hours the dye-bath will exhaust completely.

Black.—Prepare the dye-bath with 5 lb. Oxydiamine black A, 20 lb. Glauber's salt, 2 lb. soda. Dye at the boil for one hour.

Black.—Prepare the dye-bath with 6 lb. Pluto black B. Dye at the boil for one hour with the addition of ¾ to 1¼ oz. Glauber's salt, ½ to ¾ oz. soda ash per gallon of liquor. To develop the shade it is necessary to dye in a boiling liquor.

Black.—Use 2½ lb. Diamine jet black Cr, 2½ lb. Diamine jet black R B, 2 lb. soda, and 20 lb. Glauber's salt, working at the boil for one hour.

Black.—Use 6 lb. Oxydiamine black N R, 2 lb. soda and 20 lb. Glauber's salt, working at the boil for one hour.

Black.—Use 6 lb. Columbia Black F B B, 3 lb. soda and 20 lb. Glauber's salt, working at the boil to shade.

Besides the blacks given in the above recipes, there are other brands which are used in the same way, and vary slightly in the shade of black they give.

All the direct blacks require working in strong baths to give anything like black shades; they all have, more or less, a bluish tone, which can be changed to a jetter shade by the addition of a yellow or green dye in small proportions, which has been done in one of the recipes given above.

By coupling, chroming or developing, the direct blacks can be made to give full, deep and fast blacks, and examples of their use in this manner will be found in following sections.

By using all the direct blacks in proportions varying from ¼ to 1 per cent. of dye-stuff to the weight of the cotton they give greys of various tints and depths; a few examples of such greys will now be given.

Blue Grey. — Prepare the dye-bath with ½ lb. Oxydiamine black A, 1 lb. soda, 10 lb. Glauber's salt. Dye at the boil for one hour. [Pg 106]

Bright Grey. — Prepare a dye-bath with 4½ oz. Azo mauve A M, 1½ oz. Direct yellow G, 3 lb. soda, 15 lb. common salt.

Silver Grey. — Prepare the dye-bath with ½ oz. Neutral grey G, 10 lb. sulphate of soda. Work at the boil to shade, then lift, wash and dry.

Slate. — Dye in a bath with ½ lb. Diamine black B H, 3 oz. Diamine bronze G, 15 lb. Glauber's salt at the boil for three-quarters of an hour.

Bronze Grey. — Prepare a dye-bath with ½ lb. Diamine bronze G, 15 lb. Glauber's salt, 3 lb. soap. Enter at about 160° F., raise to boil and work for one hour; lift, wash and dry.

Dark Slate. — Prepare a dye-bath with 10 lb. Glauber's salt, 1½ lb. soap, 1 lb. Diamine black R O, 2 lb. Cotton brown N. Heat to about 150° F. Enter the goods, work for a short time, then raise to the boil and work for one hour; lift, wash and dry.

Green Grey. — Prepare a dye-bath with 10 lb. Glauber's salt, 1 lb. Diamine black R O, ¾ oz. Thioflavine S. Enter at from 150° Tw. to 180° F., raise to boil and dye for one hour; wash and dry.

Light Slate. — Prepare a dye-bath containing 2¼ lb. soap, 15 lb. Glauber's salt, 6 oz. Diamine black R O, ½ oz. thioflavine S. Enter cotton at 140° F., work a little, then heat to boil and dye to shade; lift, wash and dry.

Grey. — Prepare the dye-bath with ½ lb. Diamine grey G, ½ oz. Diamine scarlet B, 1 lb. soda, 1 lb. soap, 5 lb. Glauber's salt. Dye for one hour at the boil.

Light Grey. — Prepare the dye-bath with 1 lb. Diamine grey G, 1 lb. soda, 1 lb. soap, 5 lb. Glauber's salt. Dye for one hour at the boil; lift, rinse and dry.

It may be convenient here to deal with the question of the fastness of the direct dyes to such influences as light, air, [Pg 107] acids, alkalies, washing and soaping, that have a very material influence on the use of these dyes in dyeing various fabrics. This matter can only be dealt with here in very general terms, for space is limited and the dyes are too numerous for detailed mention. They vary very greatly in degrees of fastness, some are absolutely fast to all influences; the blacks are among the fastest, generally these resist washing and soaping, stand acids well and are fast to alkalies, light however affects them more or less, though they cannot be reckoned fugitive colours. The few direct greens known are good colours; they stand washing, soaping and light well, but they are affected by acids and alkalies. The blues vary very much, generally they stand soaping and have a fair degree of fastness to light, acids have but little action, alkalies tend to redden the shade, while heat also affects them. The direct browns are very variable; they are in general not fast to light; they stand washing and soaping and resist alkalies, but are altered by acids slightly. The yellows rank among the fastest of colours to light and washing and soaping; acids have but little effect; they are reddened by alkalies. Among the reds there is great variation in properties, generally they are not fast to light, standing washing and soaping well and resisting weak alkalies; some of them, such as the Benzo purpurines and Congo reds are very sensitive to acids, being turned blue with very weak acids, but on washing or soaping the original colour comes back; others, like the Titan reds, Diamine reds and Delta purpurines are not so sensitive, but these are affected by moderately strong acids; there are one or two reds like Benzo fast scarlet 4 B S and Purpuramine D H, which are fast to acids. The depth of shade which is dyed has some considerable influence on the degree of fastness, the deeper shades of a colour are always faster than the paler shades, particularly as regards light, a difference of ½ per cent, of dye-stuff has been known to make a very [Pg 108] appreciable degree of difference as regards the fastness of a colour to light.

In dyeing cotton with all the direct dyes, it is found that the whole of the dye-stuff is not removed from the dye-bath, how much is taken up by the cotton, and the depth of the shade which is dyed upon the cotton chiefly depends upon three factors: —

(1) Volume of water used. (2) Quantity of saline salts used. (3) Degree of affinity of the dye-stuff for the fibre.

There may also be some minor factors such as temperature at which the dyeing is carried on, the character and condition of the fabrics being dyed, etc.

The volume of water used in making the dye-bath has a very great influence upon the amount of dye taken up by the cotton, the greater the volume of water the less dye is absorbed and the paler the colour which is produced upon the fibre. It is therefore important to use as little water as possible in making up the dye-bath, indeed, for anything like good results to be obtained with some dyes, especially those of the sulphur series like Vidal black, Immedial blacks, Katigen browns, Cross-dye blacks, Amidazol blacks, etc., it is necessary to employ what is called a short bath, that is making it as strong as possible. The proportion of water with such dyes should not exceed fifteen times the weight of the cotton being dyed, that is, for every pound of cotton, 1½ gallons of water can be allowed. This will suit the dyeing of yarns and loose fabrics like knitted stockings and hosiery goods very well. In the case of dyeing piece goods on a jigger or continuous dyeing machines even stronger liquors can be used with advantage. With some of the older, direct dyes like Congo red, Benzo azurine, Diamine scarlets, the proportion of water may be increased to twenty times the weight of the [Pg 109] cotton. In any case the quantity of water used should not exceed twenty-five times the weight of the cotton.

The second factor, the quantity of saline salts, like Glauber's salt, soda, borax, etc., added in the dyeing, is not without influence, generally the more that is added the more dye there is left in the bath, but here again much depends upon the salt and the colouring matters used. Some salts, more particularly Glauber's salt and common salt, tend to throw some dye-stuffs out of the bath, and so the more there is used of them the deeper the shade produced on the fabric. It is quite impossible, having regard to the scope of this book, to deal with this question in detail. The dyer should ascertain for himself the best salts and the best proportions of these to use with the particular dyes he is using. The recipes given above will give him some ideas on this point.

The third factor, the degree of affinity of the dye for the cotton fibre, has some influence on the depth of shade which can be dyed from any given strength of the dye-bath. There is a very considerable difference among the direct dyes in this respect. There are some which have a fair degree of affinity, while there are others which have but little affinity, and while in the former case there is little dye left in the bath, in the latter case there is a good deal. When dyeing plain shades with single dye-stuffs this is not of much moment, because if the bath be kept for further use, as will be spoken of presently, the bath may be brought up to its original strength by adding a proportionate amount of dye-stuff, but when compound shades are being dyed, using two or more dyes, then this feature has some influence, for they will not be absorbed by the fibre in the same proportion as they were put in the bath, and so when making up the dye-bath for the second lot, and adding the same proportion of dyes, the shade which is produced will not be quite the same, for the first lot of cotton in taking up the dyes in vary [Pg 110] ing quantities has altered their relative proportions, and so the bath for the second lot of cotton will actually contain more of one dye than did the first bath, and the influence of this excess of the one constituent will show itself in the shade ultimately dyed. The more lots of cotton there are dyed in the bath the greater will this influence be. The dyer must by practical experience find out for himself in what direction this feature of the direct dyes exerts its influence on the particular dyes he is working with and make due allowance.

It is found in practice that from one-fourth to one-half of the original weight of dye-stuff is left in the bath, and in order to be as economical as possible a custom has arisen of keeping the bath and using it again for dyeing further lots of cotton. In thus making a continuous use of dye-baths it is important in preparing the baths for the next lot of cotton to add first the requisite quantities of dye-stuffs, how much will depend upon the factors and conditions already detailed, but from one-half to three-fourths of the original quantities are added. Practical experience alone is the guide to be followed.

Having added the dye-stuff, then sufficient water must be added to bring up the volume of the bath to the proper amount, for it will have lost some. The loss of water arises from two sources: first there is the evaporation, which always occurs when dye-baths are heated up, and, second, there is the mechanical loss due to its absorption by the material which is being dyed. When a piece of cotton or other textile fabric is immersed in a dye liquor it absorbs mechanically some of it, and this amount may be roughly put down as about its own weight; thus 100 lb. weight of cotton will take up 10 gallons of liquor and carry that quantity out of the bath. To some extent this may be minimised by a previous wetting out of the cotton, which will then have in it as much liquor as it will take up, and so practically no more will be taken up from the [Pg 111] dye-bath. Any loss of volume which may thus occur can be remedied by the addition of water.

The dye-baths containing in solution, in addition to the dye-stuff, salt, or Glauber's salt, or any other added substance, the cotton in taking up the dye liquor will of course take up some of these in proportion to the volume of liquor absorbed. The amount may range from 4 oz. to 1 lb. per gallon of liquor, and if 100 lb. cotton is being dyed and takes up from 10 to 15 gallons of liquor, it is obvious that it must absorb from 3 to 10 lb. of saline matter, and as the salinity of the dye liquor is of some importance in dyeing direct colours, in making up the bath for the next lot of cotton this must be allowed for and suitable additions made. In order to do this properly it is a good plan to rely upon the Twaddell.

The dyer should take the Twaddell of his bath before use and always make up his baths to that strength. This will be found to range from 3° to 12° Tw.

Thus, for instance, a dye-bath made from 120 gallons of water with 20 lb. to 25 lb. common salt or Glauber's salt with the dye-stuffs will stand at 4° Tw., one made with 50 lb. common salt or Glauber's salt at 8° Tw., while one which is made with 80 lb. to 100 lb. salt will stand at 12° to 13° Tw. If the dyer always maintains his liquors at one uniform degree Twaddell he can invariably depend upon getting uniform shades from his dye-baths. This uniform strength is attained by adding more salt or more water as the case may require.

Of course the continuous working of dye-baths cannot go on for ever; sooner or later the baths become thick and dirty, and then they must be thrown away and a new bath started. [Pg 112]

(2) DIRECT DYEING FOLLOWED BY FIXATION WITH METALLIC SALTS.

It is an acknowledged principle in dyeing that to produce colours fast to washing, soaping and rubbing, there must be produced on the fibre an insoluble coloured substance. Now as the direct dyes do not essentially produce such insoluble bodies when dyed on the cotton, the colours they form are not always fast to washing and soaping. It has been ascertained, however, that some of the direct dyes, *e.g.*, Benzo azurine, Chicago blue, Catechu browns, Diamine blues, Diamine browns, etc., are capable of uniting with metallic bodies to form insoluble colour lakes, and this combination can take place on the fibre. Fast shades may be dyed with the dye-stuffs named above, and with others of this group, by first dyeing them in the usual way, then passing through a boiling bath containing bichromate of potash or copper sulphate, either together or separately. The two fixing agents here named have been found to be the best, although others, as, for instance, zinc sulphate, chromium fluoride and iron sulphate have been tried. With some dyes there is little or no alteration in shade, but in others there is some change, thus the blues as a rule tend to become greener in tone, and browns also tend to acquire a greener tone and deeper shade. The treated shades thus obtained are notable for considerable fastness to washing, soaping and light. It is to be noted that bichromate of potash exercises both a fixing and an oxidising action on dye-stuffs, hence it is needful to use it with some degree of caution and not in too great an amount, otherwise with some dyes there is a risk of over-oxidation, and in consequence poor shades will be developed. The following recipes will serve to show what dyes may thus be used, and the colours that can be obtained with them. [Pg 113]

Dark Red.—Use in the dye-bath 3 lb. Diamine fast red F, 3 lb. soda and 20 lb. Glauber's salt, work at the boil for one hour, then lift, rinse and pass into a boiling bath containing 3 lb. fluoride of chromium for ten to fifteen minutes, then lift, rinse and dry. By using 1 lb. of the dye-stuff in the same way a light red shade is got.

Orange.—Dye at the boil for one hour with 1 lb. Chrysamine G, 3 lb. soap and 10 lb. Glauber's salt, then rinse and fix in a fresh boiling bath with 1 lb. bichromate of potash, 3 lb. sulphate of copper and 2 lb. acetic acid.

Yellow.—Dye with 3½ lb. Diamine yellow N, 3 lb. soap and 15 lb. phosphate of soda, then fix with 4 lb. fluoride of chromium.

Gold Yellow.—Prepare the dye-bath with 3 lb. Benzo chrome brown 5 G, 1 lb. soda ash, 12 lb. Glauber's salt. Dye at the boil for one hour and rinse. This gives an orange brown. To get the yellow shade, afterwards chrome with 3 lb. bichromate of potash, 3 lb. sulphate of copper, 1 lb. acetic acid, in a fresh bath. Enter at about 130° F., bring to the boil, and boil for half an hour.

Pale Leaf Green.—Dye with 3 lb. Dianil yellow 3 G, 1 lb. Dianil yellow R, 1 lb. Dianil blue G, and 20 lb. salt, then fix with 3 lb. copper sulphate and 2 lb. acetic acid.

Leaf Green.—Dye with 3 lb. Dianil yellow 3 G, 3 lb. Dianil blue G, and 20 lb. salt, fixing with 4 lb. copper sulphate and 2 lb. acetic acid.

Dark Green.—Dye with 2 lb. Dianil yellow R, 1½ lb. Dianil dark blue R, 1 lb. soda and 20 lb. salt, fixing with 3 lb. copper sulphate.

Pale Olive Green.—Dye with 2¾ lb. Diamine fast yellow B, 1¼ lb. Diamine blue R W, ¾ lb. Diamine blue R W, ¾ lb. Diamine catechine G; fix with 4 lb. sulphate of copper and 2 lb. acetic acid.

Russia Green.—Dye with 2½ lb. Diamine blue R W, 10 oz. [Pg 114] Diamine dark blue B, 2½ lb. Diamine fast yellow B, 3 lb. soda and 20 lb. Glauber's salt; fix with 4 lb. sulphate of copper and 2 lb. acetic acid.

Blue Green.—Dye with 1¾ lb. Diamine sky blue F F, 6 oz. Diamine fast yellow B, 1 lb. soda and 10 lb. Glauber's salt; fix with 2 lb. sulphate of copper and 1 lb. acetic acid.

Bronze Green.—Use in the bath at the boil 4 lb. Diamine bronze G, 2 lb. soda and 10 lb. Glauber's salt, then fix with 4 lb. fluoride of chromium.

Pea Green.—Dye in a boiling bath with ½ lb. Diamine sky blue F F, 2¼ lb. Diamine fast yellow A, 1 lb. soda and 10 lb. Glauber's salt, then fix in a fresh bath with 2 lb. sulphate of copper and 1 lb. acetic acid.

Leaf Green.—Dye at the boil for one hour in a bath containing 2¾ lb. Diamine fast yellow B, 1¾ lb. Diamine blue R W, 7 oz. Diamine catechine B, 2 lb. soda and 20 lb. Glauber's salt, then fix in a new bath with 4 lb. sulphate of copper and 2 lb. acetic acid.

Light Green.—Prepare the dye-bath with 7¼ oz. Diamine blue R W, 5½ oz. Diamine orange B, 2 lb. Diamine fast yellow B, 1 lb. soda and 10 lb. Glauber's salt, work at the boil for one hour, then treat in a fresh bath with 3 lb. sulphate of copper.

Olive Green.—Dye with 2¼ lb. Chicago blue R W, 15 oz. Chrysamine G, 2 lb. soda and 10 lb. Glauber's salt; fix with 1 lb. bichromate of potash, 3 lb. sulphate of copper and 2 lb. acetic acid.

Pea Green.—Use in the dye-bath 3 lb. Chrysophenine G, 1 lb. Chicago blue 6 B, 2 lb. soda and 10 lb. Glauber's salt, working at the boil for one hour, then fix in a fresh boiling bath with 3 lb. sulphate of copper and 2 lb. acetic acid.

Green.—Dye with 2¾ lb. Chicago blue 6 B, 5 oz. Chrysamine G, 2 lb. soap and 20 lb. Glauber's salt; fix with 1 lb. bichromate of potash, 3 lb. sulphate of copper and 2 lb. acetic acid. [Pg 115]

Dark Green.—Dye with 1½ lb. Diamine green B, 1½ oz. Diamine bronze G, 1 lb. Diamine fast yellow A, 3 lb. soda and 20 lb. Glauber's salt, working at the boil for one hour, then lift, rinse and fix in a fresh boiling bath with 3 lb. fluoride of chromium for one to fifteen minutes.

Dark Bronze.—Use in the dye-bath 2½ lb. Diamine bronze G, 3 lb. soda and 20 lb. Glauber's salt, working at the boil for one hour, then lift, rinse and fix with 3 lb. fluoride of chromium as above.

Dark Blue.—Prepare the dye-bath with 3 lb. Benzo blue R W, 10 lb. Glauber's salt; dye for one hour at the boil, then treat in fresh bath with 1 lb. sulphate of copper at the boil for half an hour.

Blue.—Dye with 1¾ lb. Diamine brilliant blue G, 1¼ lb. Diamine sky blue F F, 2 lb. soda and 20 lb. Glauber's salt; fix in a bath with 4 lb. sulphate of copper and 2 lb. acetic acid.

Light Navy.—Dye with 1 lb. Diamine blue 3 R, 2¼ lb. Diamine blue R W, 2 lb. soda and 20 lb. Glauber's salt; fix with 4 lb. sulphate of copper and 20 lb. acetic acid.

Bright Navy.—Dye with 4 lb. Diamine brilliant blue G, 2 lb. soda and 20 lb. Glauber's salt; fix with 4 lb. sulphate of copper and 2 lb. acetic acid.

Blue.—Dye with 3 lb. Chicago blue R W, 3 lb. soda and 20 lb. Glauber's salt; fix with 3 lb. sulphate of copper and 2 lb. acetic acid.

Dark Blue.—- Dye with 3 lb. Chicago blue R W, 1½ lb. Zambesi black F, 3 lb. soda and 20 lb. Glauber's salt; fix with 3 lb. sulphate of copper and 2 lb. acetic acid.

Deep Slate Blue.—Dye with 1¼ lb. Zambesi black F, 1¼ lb. Chicago blue B, 6 oz. Columbia yellow, 3 lb. soda and 20 lb. Glauber's salt; fix with 3 lb. sulphate of copper and 2 lb. acetic acid.

Light Blue.—Prepare the dye-bath with 2 oz. Diamine [Pg 116] sky blue F F, ¾ oz. Diamine fast yellow A, ½ lb. soda, 2 lb. soap and 5 lb. Glauber's salt; dye for one hour at the boil, then treat in a fresh bath with 1½ lb. sulphate of copper for half an hour.

Dark Blue.—Prepare the dye-bath with 4 lb. Benzo chrome black blue B, 15 lb. Glauber's salt and 3 lb. soda. Work at the boil for one hour, then chrome in a fresh bath with 1 lb. bichromate of potash, 1 lb. sulphate of copper and ½ lb. sulphuric acid.

Dark Blue.—Dye with 2½ lb. Diamineral blue R, 3 lb. soda and 20 lb. Glauber's salt; fix with 2 lb. sulphate of copper, 2 lb. bichromate of potash and 2 lb. acetic acid.

Turquoise Blue.—Dye with 1 lb. Chicago blue 6 B, 2 lb. soda and 10 lb. Glauber's salt, and fix with 3 lb. sulphate of copper and 2 lb. acetic acid.

Dark Turquoise Blue.—Dye with 3 lb. Chicago blue 4 B, 2 lb. soda and 10 lb. Glauber's salt, and fix with 3 lb. sulphate of copper and 2 lbs. acetic acid.

Black Blue.—Dye with 4¼ lb. Diamine dark blue B, 1 lb. Diamine new blue R, 2 lb. soda and 10 lb. Glauber's salt, fixing with 5 lb. sulphate of copper and 2 lb. acetic acid.

By mixing together the various Diamine blues a very great range of shades can be produced, from pale sky-blue tints to the deepest of blues.

Bright Blue.—Dye with 2¼ lb. Dianil blue B and 20 lb. Glauber's salt; fix with 3 lb. of fluoride of chromium.

Dark Blue.—Dye with 3 lb. Dianil blue B, 1 lb. Dianil dark blue R, 1 lb. soda and 20 lb. salt, fixing with 3 lb. fluoride of chromium.

Red Violet.—Dye with 1 lb. Dianil blue 4 R and 10 lb, salt, fixing with 4 lb. fluoride of chromium.

Dark Plum.—Dye with 3 lb. Dianil blue 4 R and 15 lb. salt, fixing with 4 lb. fluoride of chromium.

Red Violet.—Dye with 1 lb. Diamine blue 3 R, 1 lb. soda [Pg 117] and 10 lb. Glauber's salt, fixing with 1½ lb. sulphate of copper and 1 lb. acetic acid.

Red Plum.—Use 3¾ lb. Diamine blue 3 R, 3 lb. soda and 20 lb. Glauber's salt, fixing with 5 lb. sulphate of copper and 2 lb. acetic acid.

Dark Brown.—Prepare the dye-bath with 5 lb. Diamine catechine B, 3 lb. soda and 15 lb. Glauber's salt and dye at the boil for one hour, then treat with 2 lb. sulphate of copper and 2 lb. bichromate of potash.

Brown.—Prepare the dye-bath with 4 lb. Paramine brown C, 20 lb. Glauber's salt, 2 lb. soda and dye at the boil for one hour; treat with 3 lb. copper sulphate.

Light Brown.—Dye at the boil for one hour in a bath containing 5 lb. Diamine catechine G, 3 lb. soda and 15 lb. Glauber's salt, then treat in a fresh bath with 2 lb. sulphate of copper and 2 lb. bichromate of potash.

Dark Chestnut Brown.—Dye for an hour in a boiling bath with 2¼ lb. Diamine catechine G, 1¼ lb. Diamine fast yellow B, 3 lb. soda and 20 lb. Glauber's salt; then fix in a fresh boiling bath with 2 lb. sulphate of copper, 2 lb. bichromate of potash and 2 lb. acetic acid, working for fifteen to twenty minutes, then rinsing and drying.

Brown.—Use 3 lb. Catechu brown G K, 15 lb. Glauber's salt and ½ lb. soap; after dyeing for one hour at the boil treat in a fresh boiling bath with 3 lb. copper sulphate.

Dark Brown.—Dye at the boil for one hour with 3 lb. Catechu brown F K, 15 lb. Glauber's salt and 1 lb. soap, then treat in a fresh boiling bath with 3 lb. copper sulphate.

Brown.—Prepare the dye-bath with 9 oz. Diamine blue R W, 12½ oz. Diamine orange B, 1¾ lb. Diamine fast yellow B, 2 lb. soda and 20 lb. Glauber's salt; after working for one hour at the boil treat in a fresh boiling bath with 4 lb. sulphate of copper.

Brown.—Prepare the dye-bath with 4 lb. Benzo chrome [Pg 118] brown 2 R, 20 lb. Glauber's salt (crystals) and dye at the boil for one hour; afterwards treat with bichromate of potash and sulphate of copper.

Nut Brown.—Dye in a bath with 4 lb. Benzo chrome brown G and 20 lb. salt, then treat in a fresh bath with 4 lb. bichromate of potash, 4 lb. copper sulphate and 1 lb. acetic acid.

Chestnut Brown.—Dye at the boil for one hour in a bath containing 4 lb. Benzo chrome brown R, and boiling bath with 4 lb. bichromate of potash, 4 lb. sulphate of copper and 1 lb. acetic acid.

Dark Olive Brown.—Dye with 4 lb. Diamine bronze G, 1 lb. Diamine orange B, 2 lb. soda and 20 lb. Glauber's salt; fix with 5 lb. sulphate of copper and 2 lb. acetic acid.

Deep Brown.—Use in the Dye-bath 1¾ lb. Diamine brown B, 1¾ lb. Diamine fast yellow B, ½ oz. Diamine black B H, 3 lb. soda and 20 lb. Glauber's salt. The fixing bath contains 2 lb. sulphate of copper, 2 lb. bichromate of potash, and 2 lb. acetic acid.

Dark Brown.—Dye with 2 lb. Diamine brown M, 1 lb. Diamine fast red F, ½ lb. Diamine jet black Cr, 3 lb. soda and 20 lb. Glauber's salt. The fixing bath contains 2 lb. sulphate of copper, 2 lb. bichromate of potash and 2 lb. acetic acid.

Black Brown.—Dye with 1¾ lb. Diamine dark blue B, ¾ lb. Diamine orange B, 1¾ lb. Diamine fast yellow B, 2 lb. soda and 20 lb. Glauber's salt, fixing with 5 lb. sulphate of copper and 2 lb. acetic acid.

Light Sage Brown.—Dye with ¾ lb. Diamine brown B, 1½ lb. Diamine fast yellow B, 3 oz. Diamine dark blue B, 2 lb. soda and 20 lb.

Glauber's salt, fixing with 3 lb. sulphate of copper and 1 lb. acetic acid.

Pale Brown.—Use in the dye-bath 1 lb. Dianil brown 3 G O, 4 oz. Dianil brown E, 4 oz. Dianil black N, 1 lb. soda and 20 [Pg 119] lb. salt, fixing with 1½ lb. sulphate of copper and 1 lb. acetic acid.

Walnut Brown.—Dye with 2½ lb. Diamine blue 3 R, 1 lb. Diamine brown M, 2 lb. soda and 20 lb. Glauber's salt, then fix with 5 lb. sulphate of copper and 2 lb. acetic acid.

Pale Fawn Brown.—Dye with 2 lb. Diamine blue 3 R, 1 lb. Diamine brown M, 2 lb. soda and 20 lb. Glauber's salt, then fix with 5 lb. sulphate of copper and 2 lb. acetic acid.

Pale Fawn Brown.—Dye with ½ lb. Diamine orange B, ¼ lb. Diamine fast yellow B, 1 lb. soda and 10 lb. Glauber's salt, fixing with 2 lb. sulphate of copper and 1 lb. acetic acid.

Sage Brown.—Dye with 9 oz. Diamine blue R W, ¾ lb. Diamine orange B, 1¾ lb. Diamine fast yellow B, 2 lb. soda and 20 lb. Glauber's salt. The fixing is done with 4 lb. sulphate of copper and 2 lb. acetic acid.

Red Chocolate.—Dye with 3 lb. Diamine orange B, 1 lb. soda and 10 lb. Glauber's salt; fix with 2 lb. sulphate of copper and 1 lb. acetic acid.

Dark Chestnut.—Dye with 2½ lb. Dianil brown 3 G O, 13 oz. Dianil brown R, 13 oz. Dianil brown B D, 1 lb. soda and 20 lb. salt, fixing with 3 lb. copper sulphate and 1 lb. acetic acid.

Brown.—Dye with 2¼ lb. Chrysophenine G, 1¼ lb. Diamine brown G, 1¼ lb. Chicago blue R W, 3 lb. soda and 20 lb. Glauber's salt; fix with 3 lb. sulphate of copper and 2 lb. acetic acid.

Nut Brown.—Dye with 3 lb. Chromanil brown 2 G, 3 lb. soda and 20 lb. Glauber's salt; fix with 1 lb. bichromate of potash, 3 lb. sulphate of copper and 2 lb. acetic acid.

Dark Grey.—Dye at the boil for one hour with 1 lb. Zambesi black F, 3 lb. soda and 10 lb. Glauber's salt; fix in a fresh boiling bath with 3 lb. sulphate of copper, 1 lb. bichromate of potash and 10 lb. Glauber's salt.

Dark Grey. — Dye with 3 lb. Chromanil black 4 R F, 3 lb. soda and 10 lb. Glauber's salt; fix with 1 lb. bichromate of potash, 3 lb. sulphate of copper and 2 lb. acetic acid. [Pg 120]

Dark Grey. — Use in the dye-bath 1 lb. Diamine blue R W, ½ lb. Diamine orange B, ¼ lb. Diamine new blue R, 2 lb. soda and 20 lb. Glauber's salt, fixing with 4 lb. sulphate of copper and 2 lb. acetic acid.

Pale Greenish Grey. — Dye with ¼ oz. Diamine orange B, 3 oz. Diamine blue R W, ½ lb. soda, 2 lb. soap and 5 lb. Glauber's salt, fixing with 1 lb. sulphate of copper and ½ lb. acetic acid.

Slate Blue. — Dye with ¼ lb. Diamine dark blue B, 2 oz. Diamine new blue R, 1 lb. soda and 10 lb. Glauber's salt; fix with 2 lb. sulphate of copper and 1 lb. acetic acid.

Grey. — Prepare the dye-bath with 2 lb. Cross-dye black 2 B, 5 lb. soda ash, 15 lb. common salt; after rinsing leave the cotton in the air to age overnight, rinse again and work for half to three-quarters of an hour at from 150° to 160° F. in a bath containing 5 lb. bichromate of potash and 5 lb. sulphuric acid, then thoroughly rinse and dry.

Dark Grey. — Dye with 1 lb. Diamine jet black Cr, 1 lb. soda and 10 lb. Glauber's salt, fixing with 1 lb. bichromate of potash and ½ lb. acetic acid.

Green Grey. — Dye with 1 lb. Diamine dark blue B, 2 oz. Diamine orange B, 4 oz. Diamine fast yellow B, 1 lb. soda and 10 lb. Glauber's salt, fixing with 3 lb. sulphate of copper and 1 lb. acetic acid.

Grey. — Dye with 4 oz. Dianil black N, 1 lb. soda and 10 lb. salt, fixing with 1 lb. copper sulphate and ½ lb. acetic acid.

Black. — Prepare the dye-bath with 5½ lb. Diamine jet black R B, 1 lb. Diamine dark blue B, 20 lb. Glauber's salt; dye at the boil for one hour, rinse and then treat the goods simmering for twenty minutes with 4 lb. bichromate of potash.

Black. — Prepare the dye-bath with 8 lb. Chromanil black R F and 20 lb. Glauber's salt; dye at the boil for one hour, then treat boiling hot for about thirty minutes in a fresh bath [Pg 121] with 1 lb. bichromate of potash and 3 lb. sulphate of copper. Add 6 lb. only of the dye-stuff to the bath for a second batch.

Black. — Use 5 lb. Dianil black N, 5 lb. soda and 20 lb. salt; then fix with 3 lb. copper sulphate, 3 lb. bichromate of potash and 2 lb. acetic acid.

Black. — Use in the dye-bath 5 lb. Dianil black C R. 3 lb. caustic soda, 36° Tw. and 20 lb. salt, fixing with 3 lb. copper sulphate, 3 lb. bichromate of potash and 2 lb. acetic acid.

Jet Black. — Dye with 5 lb. Diamine jet black Cr, 1 lb. soda and 20 lb. Glauber's salt, fixing with 4 lb. bichromate of potash and 2 lb. acetic acid.

It will be convenient here to deal with a small but growing and important class of dye-stuffs which contain sulphur in their composition, and which, therefore, are named: —

Sulphur or Sulphyl Colours.

The original type of this group is Cachou de laval, sent out a good many years ago, but of late years Vidal black, St. Dennis black, Cross-dye blacks and drab, Immedial blacks, blues and browns, Amidazol blacks, browns and olives, Sulfaniline black and brown, Katigen blacks, greens and browns, etc., have been added, and the group is likely to become a very numerous one in the future.

All these colours are dyed on to the cotton or linen from baths containing soda and salt, while some require the addition of sodium sulphide or caustic soda in order to have the dye-stuff properly dissolved. They are very weak dyes compared with the direct colours, and require from 20 to 60 per cent. to produce full shades, although of this fully one-third remains in the bath unabsorbed by the cotton. It is, therefore, important in order to work as economically as possible to retain the bath, bringing it up to strength by the addition of fresh dye-stuffs, etc. [Pg 122]

Most of the dyes require the dyed goods to pass through a second bath of some reagent, bichromate of potash, sulphate of copper, etc., in order to fully develop and fix the dye on the fabric.

The best method of using the various dyes of this group will be given in the form of formulæ. Two points of importance are to use as strong a dye liquor as possible, and to expose the cotton as little as possible to the air during the dyeing operation. The dye-stuffs when exposed to the air readily become oxidised, and are thereby converted into insoluble products which become fixed on the fibre in a loose form, and in that case the dyed fibre rubs rather badly.

Pale Brown.—Prepare a dye-bath with 15 lb. Cachou de laval, 10 lb. of soda, and 10 lb. salt. The bath is not exhausted of colouring matter, and by adding one-half of the above quantities of dye-stuff and salt may be used again for another lot of cotton. After the dyeing the cotton is passed into a fixing bath of 2 lb. bichromate of potash and 1 lb. acetic acid, working at 180° F. ten to fifteen minutes.

Black.—Prepare the dye-bath with 200 gallons of water, 10 lb. soda, 10 lb. sulphide of sodium, 60 lb. salt and 16 lb. Immedial black V extra. Work at the boil for one hour, keeping the cotton well under the surface during the operation, in the case of yarns this is effected by using bent iron rods on which to hang the hanks in the vat, in the case of pieces by working with vats the guide rollers of which are below the surface of the dye liquor. After the dyeing the yarn or pieces are squeezed, well rinsed in water, then passed into the fixing bath, which contains 2 lb. sulphate of copper, 2 lb. bichromate of potash and 3 lb. of acetic acid, for half an hour at 170° to 180° F. Bichromate of potash used alone gives a reddish shade of black, sulphate of copper a greenish shade, a mixture of the two gives a greenish shade.

There are three brands of Immedial black, *viz.,* V extra, G [Pg 123] extra and F F, which vary a little in the tone of black they produce. The method of using is identical for all three. The dye-bath is not exhausted of colour and so should be kept standing, for each subsequent lot of cotton add 8 lb. Immedial black and 3 lb. sulphide of soda, and to every 10 gallons of water added to bring the bath up to volume ½ lb. soda and 3 lb. salt.

These blacks are very fast to washing, light, etc. By using smaller quantities of dye-stuff good greys can be dyed.

Black.—Prepare the dye-bath with 10 lb. soda, 10 lb. sulphide of sodium, 60 lb. salt and 25 lb. Vidal black, work at the boil for one hour, then rinse and fix with 3 lb bichromate of potash and 2 lb. sulphuric acid.

Black.—Prepare the dye-bath with 30 lb. Cross-dye black B, 10 lb. soda, 150 lb. salt. Dissolve the dye-stuff in boiling water, then add the soda crystals and finally the salt. Enter the previously well-boiled cotton at about 175° F. After a few turns raise the temperature to the boil as quickly as possible, and work for one hour (just at the boil). Lift and thoroughly rinse without delay. (The better the cotton is washed the clearer the ultimate shade.) After washing, wring up and let air age for about one hour; the intensity of the black is thereby increased.

Meanwhile prepare a bath with 5 lb. bichromate of potash, 4 lb. sulphuric acid (168° Tw.). Enter at 150° to 160° F., and work at this for about ten minutes. After chroming, wash thoroughly to remove all traces of acid. At this stage, the usual softening may take place if desirable, and finally dry at a low temperature.

The bath is kept up for further lots, and three-fourths the quantity of colouring matter, and about half soda and one fourth salt are used. Wood, or iron cisterns are most suitable, and copper pans or pipes must be avoided.

The dye-bath should be kept as short as possible, about [Pg 124] twelve to fifteen times the amount of water on the weight of cotton is advisable. The cotton when in the dye-bath should be exposed as little as possible to the air.

There are several brands of these Cross-dye blacks varying in the tone of black they give.

Black.—Prepare the dye-bath with 5 lb. soda ash, 200 lb. salt and 20 lb. Amidazol black G, this is heated to 150° F., the cotton is entered, the heat raised to the boil, and the dyeing done for an hour at that heat. Lift, rinse well, then pass into a chroming bath, made from 5 lb. bichromate of potash and 3 lb. sulphuric acid, used at 160° F. for twenty minutes, then lift, wash well and dry. The bath may be kept standing and used for other lots of cotton by replenishing with about two-thirds of the original weight of dye-stuff and a little soda. There are four brands of these Amidazol blacks which dye from a jet black with the G to a deep blue black with the 6 G brand. The G, 2 G, and 4 G, used in small quantities, 2½ to 3 lb., dye good greys of a bluish tone, the 6 G gives a dull blue, the 4 G and 6 G, used in the proportions of 7½ to 10 per cent., give dark blues.

All these blacks may be combined with aniline black with good results as shown in the following recipe: —

Black. — Prepare the dye-bath with 10 lb. Amidazol black 2 G, 5 lb. soda and 100 lb. salt. Work at the boil for an hour, then rinse, pass into a cold bath made from 2½ lb. aniline oil, 2½ lb. hydrochloric acid, 6½ lb. sulphuric acid, 7½ lb. bichromate of potash, and 5½ lb. perchloride of iron, 66° Tw. This is used cold for an hour, then the heat is slowly raised to 160° F., when the operation is finished, and the cotton is taken out well rinsed and finished as usual. Any of this class of black may be so topped with aniline black if thought necessary A very fast black is thus got.

Black. — Make the dye-bath with 15 lb. Sulfaniline black G, 60 lb. salt, 10 lb. soda, and 5 lb. sulphide of sodium. Work [Pg 125] at a little under the boil, then lift, rinse well and pass into a hot bath of 3 lb. bichromate of potash, 3 lb. sulphate of copper, and 4 lb. acetic acid for half an hour, then lift, rinse well and dry.

It has been observed in the practical application on a large scale of these sulphur blacks that the cotton is liable to become tendered on being stored, although there are few signs of such after the dyeing is finished. The exact cause of this is somewhat uncertain, the most probable reason is that during the process of dyeing a deposit of sulphur in a fine state of division has been thrown down on the cotton by decomposition of the dye-stuff, and that this sulphur has in time become oxidised to sulphuric acid which then exerts its well-known tendering action on the cotton.

The remedy for this evil lies partly with the dye manufacturer and chiefly with the dyer. The dye manufacturer should see that his product is made as free from sulphur as possible, while the dyer by careful attention to thorough washing, thorough fixation in the chrome, etc. baths, tends to eliminate all sulphur from the goods, and so prevent all possibility of the cotton becoming affected.

Blue. — Make the dye-bath with 22 lb. Immedial blue C, 13 lb. sulphide of sodium, 50 lb. salt and 15 lb. caustic soda lye at 70° Tw. Work at just under the boil for one hour, keeping the goods well under the surface of the liquor. After the dyeing the goods are well rinsed in the water and then passed into a vat which contains 1 lb. peroxide of sodium and 1 lb. sulphuric acid. This is started cold, after about fifteen minutes heat slowly to about 150°, work for twenty minutes, then lift, wash and dry. For subsequent lots of cotton there only need be used 7 lb. Immedial blue C. 2 lb. sulphide of sodium, 3 lb. salt and 1½ lb. caustic soda lye at 70° Tw. The blue may also be developed by steaming with air in a suitable chest or steaming chamber. By topping [Pg 126] with ¼ lb. New methylene blue N, very bright blue shades can be dyed.

Dark Navy.—Prepare the dye-bath with 25 lb. Immedial blue C, 24 lb. sulphide of sodium, 35 lb. common salt and 12 lb. caustic soda lye, working at the boil for one hour, then rinse and develop in a bath made from 2½ lb. peroxide of sodium and 2½ lb. sulphuric acid, started cold, then after twenty minutes heated to 160° F., twenty minutes longer at that heat will be sufficient. For second and subsequent lots of cotton there is added to the old bath 15 lb. Immedial blue C, 4 lb. sulphide of sodium, 5 lb. salt and 2 lb. caustic soda lye of 70° Tw.

Blue.—A pale but not very bright shade of blue is dyed in a bath of 3 lb. Amidazol black 6 G, 5 lb. soda and 25 lb. salt. After working for one hour at the boil, lift, rinse and pass into a bath which contains 2½ lb. peroxide of sodium and 2½ lb. sulphuric acid; this is started cold, then heated to 150° F., and kept at that heat for twenty minutes, when the cotton is taken out, well washed and dried.

Deep Blue.—Dye with 20 lb. Amidazol black 6 G, 5 lb. soda and 200 lb. salt; develop with 2 lb. peroxide of sodium and 2½ lb. sulphuric acid, working as noted above.

Dark Drab.—Prepare the dye-bath with 20 lb. Cross-dye drab, 5 lb. soda crystals and 80 lb. salt, work at the boil for an hour, then lift, wash well and dry; this can be chromed if desired.

Brown.—Dye with 20 lb. Amidazol cutch, 5 lb. soda ash and 150 lb. salt, working at the boil for one hour, then lift, wash thoroughly and dry. By after treatment in a bath of 3 lb. potassium bichromate and 3 lb. sulphuric acid the colour is made fast to washing. The shade is not altered.

Buff.—Dye with 2½ lb. Amidazol cutch, 5 lb. soda and 25 lb. salt, working at the boil for one hour, then lift, wash and dry. [Pg 127]

Pale Sea Green.—Dye with 4 lb. Amidazol green Y, 5 lb. soda and 25 lb. salt, working at the boil for one hour, then lift, wash well and dry.

Dark Green.—Dye with 20 lb. Amidazol green B, 5 lb. soda and 20 lb. salt; work at the boil for one hour, then lift, wash thoroughly and dry.

Dark Brown.—Dye with 20 lb. Amidazol cachou, 5 lb. soda and 200 lb. salt, working for an hour at the boil, then lift, rinse well and pass into a chrome bath of 4 lb. potassium bichromate and 3 lb. sulphuric acid at 50° F. for half an hour, then wash well and dry.

Dark Sage.—Dye with 20 lb. Amidazol drab, 5 lb. soda ash and 150 lb. salt for an hour at the boil, then lift and chrome with 4 lb. potassium bichromate and 8 lb. sulphuric acid for thirty minutes at 150° F., washing well afterwards.

All the Amidazol dyes are very fast to washing, acids, etc. They can be treated with sulphate of copper or peroxide of sodium when they produce good shades. They may even be diazotised and developed with beta-naphthol and phenylene diamine. The pale tints got by using from 2 to 4 per cent. of dye-stuff are useful ones, as also are the medium shades with 10 per cent. of dye-stuff.

Brown.—Prepare the dye-bath with 10 lb. Sulfaniline brown 4 B, 50 lb. salt, 10 lb. soda and 5 lb sulphide of sodium; work at the boil for one hour, then lift, wash and treat in a fresh bath with 3 lb. potassium bichromate and 2 lb. acetic acid at 160° F. for half an hour, then wash well and dry.

Olive.—Dye with 10 lb. Katigen olive G, 50 lb. salt, 10 lb. soda and 6 lb. sulphide of sodium; work for one hour at the boil, then lift, wash and treat in a fresh bath with 2 lb. bichromate of potash, 2 lb. sulphate of copper and 2 lb. acetic acid for half an hour at the boil, then wash.

Dark Olive.—Dye with 20 lb. Katigen olive G, 50 lb. salt, 10 lb. soda, and 6 lb. sulphide of sodium, working at the boil [Pg 128] for one hour, then lift, wash and dry. By chroming a darker and faster olive is got.

126

Brown.—Dye with 20 lb. Katigen dark brown, 50 lb. salt, 10 lb. soda and 6 lb. sulphide of sodium at the boil for one hour, then treat in a fresh bath with 2 lb. bichromate of potash, 2 lb. sulphate of copper and 2 lb. acetic acid for half an hour at the boil, then wash well.

Pale Brown.—Dye with 8 lb. Immedial bronze A, 2 lb. soda, 2 lb. sulphide of sodium and 10 lb. Glauber's salt at the boil for one hour, then lift, rinse and pass into a fresh bath containing 1 lb. bichromate of potash and 2 lb. acetic-acid at 150° F. for half an hour, then lift, wash and dry.

Dark Brown.—Dye with 12 lb. Immedial brown B, 5 lb. sulphide of sodium, 5 lb. soda and 20 lb. salt at the boil for one hour, then lift and treat in a fresh bath with 2 lb. bichromate of potash, 2 lb. sulphate of copper and 2 lb. acetic acid.

The Immedial blacks, blue, bronze and brown dye very fast shades, standing soaping, acids and light. They may be combined together to produce a great range of shades of blue, brown, green, grey, etc.

These examples will perhaps suffice to show how this new but important class of sulphyl colours are applied to the dyeing of cotton. They may be topped with aniline black, indigo, basic dyes, or combined with such direct dyes as produce shades fast to chroming to form a very great range of shades which have the merit of fastness.

(3) DIRECT DYEING FOLLOWED BY FIXATION WITH DEVELOPERS.

A large number of the dyes prepared from coal tar are called azo colours, such for instance are the Biebrich and Croceine scarlets and oranges, Naphthol black, Congo red, etc., [Pg 129] just to name a few. The preparation of these is about the simplest operation of colour chemistry, and consists in taking as the base an amido compound as the chemist calls such. These amido compounds, of which aniline, toluidine, benzidine, naphthylamine are familiar examples, are characterised by containing the molecular group NH_2, which radicle is built up of the two elements nitrogen and hydrogen. All compounds which contain this group are basic in character and combine with acids to form well-defined salts. When these amido bodies are treated with sodium nitrite and hydrochloric acid they undergo a chemical change, the feature of which is that the nitrogen atoms present in the amido compound and in the nitrite unite together and a new compound is produced which is called a diazo compound, and the operation is called "diazotisation".

For example when paranitroaniline is subjected to this reaction it undergoes a change indicated in the chemical equation: —

$C_6H_4NO_2NH_2$, + $NaNO_2$, + 2HCl = Paranitroaniline, Sodium nitrite, Hydrochloric acid.

$C_6H_4NO_2N$: NCl + NaCl + $2H_2O$ = Paranitro benzene Sodium chloride, Water, diazo chloride.

The above, put into words, means that when paranitroaniline is dissolved with hydrochloric acid and treated with nitrite of soda it forms diazonitro benzene chloride, sodium chloride and water. Now the diazo compounds are rather unstable bodies, but they have a great affinity for other compounds, such as naphthol, phenylene diamine, phenol, and combine easily with them when brought into contact with them. The new compounds thus made form the dye-stuffs of commerce.

The azo dyes contain the characteristic group of two [Pg 130] nitrogen atoms shown in the formula N: N. In dealing with the production of colours direct on the fibre this subject will be elaborated more fully.

Now many of the direct dyes, Diamine blacks, Diamine cutch, Primuline, Diazo brown, Zambesi blues, browns, etc., contain amido groups, by reason of having been made from such bodies as phenylene diamine, amido naphthol, toluidine, etc., and it has been found that when dyed on the fibre they are capable of being diazotised by passing the dyed fibre into a bath of sodium nitrite acidified with hydrochloric acid, and if then they are placed into a bath containing such a body as beta-naphthol, phenylene diamine, etc., new compounds or dyes are produced, which are characterised by being insoluble in water, and therefore as formed on the fibre in the manner indicated are very fast to washing, soaping and similar agencies.

Often the new or developed dye formed on the fibre differs markedly in colour from the original dye. Perhaps in no case is this more strongly shown than with Primuline. The original colour is a greenish yellow, but by using various developers, as they are called, a great variety of shade can be got, as shown in this table.

Developer. Colour produced.

Beta-naphthol Bright scarlet. Alpha-naphthol Crimson. Phenylene diamine Brown. Phenol Gold yellow. Resorcine Orange. Naphthylamine ether Blue. Blue developer A N Green.

As regards the dyeing operation, it no way differs from that described for simple direct colours. It should, however, be noted that if good results are required full shades must be [Pg 131] dyed. The cotton must be rinsed in cold water, and be quite cold before it is subjected to the diazotising operation. *Diazotising* is a simple operation, yet it must be carried out with care if good results are desired. It consists essentially in the use of an acidulated bath of sodium nitrite.

To make the bath for diazotising there is taken (for each 100 lb. of goods) sufficient water to handle them in comfortably, 8 lb. of sodium nitrite and 6 lb. hydrochloric acid. This bath must be quite cold otherwise it does not work well. The goods are handled in this for from fifteen to twenty minutes, when they are ready for the next operation. The bath is not exhausted of nitrite, etc., hence it can be kept standing, and for each succeeding lot of cotton it is strengthened up by adding one-third of the quantities of nitrite and acid originally used. Of course the bath cannot be kept for ever, sooner or later it will get dirty, and then it must be thrown away and a new bath be made up.

The diazo compounds formed on the fibre are not very stable bodies. They decompose on being exposed for any great length of time to the air, while light has a strong action on most, if not all of them; hence it follows that the diazotising process should not be carried out in a room where direct, strong sunlight can enter or fall upon the goods. Then again, after diazotising, the treated goods should not be allowed to lie about exposed to air and light, but the operation of developing should be proceeded with at once, otherwise the diazo body will decompose, and weak and defective colours are liable to be obtained on subsequent development.

For *developing*, quite a large number of substances are used. Some of these are regular articles of commerce, others are the special productions of certain firms, who advise their use with the dyes that they also manufacture. These latter are sent out under such designations as Developer B, Developer A N, or Fast-blue developer. Those most in use are beta-naphthol for red from Primuline, and for bluish blacks from [Pg 132] Diamine blacks, Diazo blacks, Zambesi blacks, etc.; for dark blues from Diamine blues, Diazo blues, etc.; for greys from Diamine blues, Neutral grey, etc. Alpha-naphthol for dark reds from Primuline, greys from Diamine blues, Neutral grey, etc. Phenylene diamine for blacks from Diamine blacks, Diazo blacks, Zambesi blacks, Triamine blacks, etc.; for dark browns from Diamine browns, Diazo browns, etc.; for light browns from Cotton browns, Diamine cutch, Primuline, etc. Naphthylamine ether for blues from Diamine blacks, etc. Phenol for claret from Diamine cutch, and for gold yellow from Primuline, etc. Resorcine for orange from Primuline, etc. Soda for browns from Diamine cutch, Diazo browns, Zambesi browns, for orange from Diamine orange, and yellow from Primuline.

Beta-naphthol.—This is by far the most important of the developers. It is a white body, insoluble in water, but readily soluble in soda lye, and a solution is easily made by taking 10 lb. beta-naphthol and heating it with 10 lb. caustic soda lye of 70° Tw. and 60 gallons of water. This bath may be used as the developing bath, or it may be diluted with more water. It is not desirable to use any more caustic soda than is necessary to dissolve the beta-naphthol, so that the bath is not too alkaline. To produce full shades it usually takes 1 per cent. of the weight of the cotton of the beta-naphthol, but it is best to use the bath as a continuous one and for the first lot of cotton use 2 per cent. of naphthol, while for each succeeding lot only 1 per cent. more naphthol need be added to the same bath.

This bath is alkaline, while the diazotising bath is acid, unless, therefore, the cotton be well washed when it is taken from the latter bath there is a risk of the alkali of the one being neutralised by the acidity of the other, and the naphthol being thrown out in an insoluble form. This, of course, is easily remedied should it occur. [Pg 133]

Developer A (Bayer) is a mixture of beta-naphthol and caustic soda in the powder form, so that a solution is obtained by simply adding water. Rather more (about 1½ per cent.) of this is required than of beta-naphthol.

Alpha-naphthol has similar properties to, and is used in the same way as, beta-naphthol; it develops much darker and rather duller colours, which are less fast to washing.

Resorcine, like naphthol, is insoluble in water, but it can be dissolved by using either soda ash or caustic soda. The latter is preferable, as the former is liable to give a developing bath that froths in working, especially if much acid has been left in the cotton from the diazotising bath. The proportions are: 10 lb. resorcine, 25 lb. caustic soda lye of 70° Tw., and 60 gallons of water; or 10 lb. resorcine, 20 lb. soda ash, and 60 gallons of water, heated until a solution is obtained. In the developing bath 1 per cent. of resorcine is usually sufficient to use. It develops an orange with Primuline.

Developer F (Bayer) is a mixture of resorcine and soda ash. It requires 1½ per cent, to make a developing bath.

Phenol, better known as carbolic acid, finds a use as a developer. It is dissolved in caustic soda, 10 lb. phenol, 15 lb. caustic soda lye of 70° Tw., and 60 gallons of water. Generally 1 per cent. is sufficient to use as a developer. It is often called yellow developer.

Naphthylamine ether is used as a developer for blues in conjunction with the Diamine blacks. It is prepared for use by dissolving in hydrochloric acid, 10 lb. naphthylamine ether powder heated with 5 lb. hydrochloric acid and 50 gallons water. About 1¼ per cent. is required to form a developing bath. Naphthylamine ether is also sent out in the form of a paste mixed with acid, and containing about 25 per cent. of the actual developer.

Fast blue developer A D (Cassella), is amidodiphenylamine. It is insoluble in water, but soluble in dilute acid, [Pg 134] 10 lb. fast blue developer A D, 5 lb. hydrochloric acid and 35 gallons of water making the bath. To develop full shades 1 to 1½ per cent, is required.

Blue developer A N (Cassella). The base of this is insoluble in water, but dissolves in soda, and is probably a naphthol-sulpho acid. The product, as met with in the market, is soluble in water, and 27 lb. dissolved in 20 gallons of water form the bath. To produce full shades 1½ per cent, is usually required.

Phenylene diamine is a most important developer. It comes into the market in two forms, as a powder, very nearly pure, made into a solution by dissolving 10 lb. with 20 gallons of water and 5 lb. hydrochloric acid, and as a solution prepared ready for use. Developer C (Bayer) and developer E (Bayer) are preparations of diamine, the former in a powder, the latter in a solution. Phenylene diamine can be used with the addition to the developing bath of acetic acid or soda.

Schaeffer's acid is a sulpho acid of beta-naphthol, and is dissolved by taking 10 lb. of the acid and 7½ lb. soda, boiling with 50 gallons of water. About 1¼ per cent. is required for developing full shades.

Developer B (Bayer) is ethyl beta-naphthylamine, in the form of its hydrochloric acid compound. The bath is made from 10 lb. of the developer and 50 gallons of water, 1¼ per cent. being used to obtain full shades.

Developer D (Bayer) is dioxy-naphthalene-sulpho acid, and simply requires dissolving in water to make the bath.

Toluylene diamine is a homologue of phenylene diamine and is used in precisely the same way.

Generally the special developers issued by the various colour firms simply require dissolving in water to form the developing bath.

The cotton, previously being passed through the diazo [Pg 135] tising bath, is then run into the developing bath, in which it is kept for from twenty to thirty minutes or until the required shade is fully developed, after which it is taken out, rinsed and dried. The method of working is the same for all the developers, and may be carried out in any kind of vessels. As is indicated above, the developing baths may be kept standing and be freshened up as required; they are used cold. Sometimes two developers are mixed together, in which case care should be taken that an alkaline developer naphthol or phenol be not mixed with an acid developer (phenylene diamine, naphthylamine, etc.), unless the acidity of the latter has been neutralised with soda; otherwise the developer might be thrown out of the bath in an insoluble and hence useless form.

The advantages of the diazotising and developing process just described may be summed as—easy and quick working, superior fastness to washing, soaping and milling, increased fastness to light and softness of the dyed fibre.

Scarlet.—Dye with 3 lb. Primuline and 20 lb. salt, at the boil for one hour, diazotise and develop with beta-naphthol.

Crimson.—Dye with 3 lb. Primuline and 20 lb. salt, then diazotise and develop with alpha-naphthol.

Red Brown.—Dye with 4 lb. Primuline and 20 lb. salt, then diazotise and develop with phenylene diamine.

Deep Orange.—Dye with 3 lb. Primuline and 20 lb. salt, then diazotise and develop with resorcine.

Pale Orange.—Dye with 3 lb. Primuline and 20 lb. salt, then diazotise and develop with phenol.

Sage Brown.—Dye with 6 lb. Primuline, 3 lb. Titan ingrain blue and 20 lb. salt, then diazotise and develop with resorcine.

Dark Maroon.—Dye with 6 lb. Primuline, 3 lb. Titan ingrain blue and 20 lb. salt, then diazotise and develop with beta-naphthol. [Pg 136]

Dark Crimson.—Dye with 5¾ lb. Primuline, ¼ lb. Titan ingrain blue and 20 lb. salt, then diazotise and develop with beta-naphthol.

Dark Blue.—Dye with 3 lb. Zambesi blue B X, 2 lb. soda and 20 lb. Glauber's salt, then diazotise and develop with amidonaphthol ether.

Dark Brown.—Dye with 8 lb. Zambesi brown 2 G, 2 lb. soda and 20 lb. Glauber's salt, then diazotise and develop with toluylene diamine.

Blue Black. — Dye with 4 lb. Zambesi blue B X, 2 lb. Zambesi black D, 2 lb. soda and 20 lb. salt, then diazotise and develop with ¾ lb. toluylene diamine and ½ lb. beta-naphthol.

Red. — Dye with 4½ lb. Primuline, ½ lb. Diamine fast yellow A and 20 lb. salt, then diazotise and develop with beta-naphthol.

Dark Brown. — Dye with 4 lb. Primuline, 1 lb. Diamine azo blue R R, and 20 lb. salt, then diazotise and develop with beta-napthol.

Deep Chestnut Brown. — Dye with 5 lb. Diamine cutch, 1 lb. soda and 20 lb. Glauber's salt, then diazotise and develop by passing for twenty minutes in a boiling bath of soda.

Dark Brown. — Dye with 4 lb. Diamine cutch, 1 lb. Diamine black B H, 2 lb. soda and 20 lb. Glauber's salt, then diazotise and develop with phenol.

Black Brown. — Dye with 1 lb. Diamine brown M, 1½ lb. Primuline, 1 oz. Diamine black B H, 2 lb. soda and 20 lb. Glauber's salt, then diazotise and develop with phenylene diamine.

Blue. — Dye with 2 lb. Diaminogene blue B B, ½ lb. soda and 20 lb. Glauber's salt, then diazotise and develop with beta-naphthol. A dark blue is got by using 8 lb. of Diaminogene blue B B in the same way.

Dark Blue. — Prepare the dye-bath with 1½ lb. Diaminogene [Pg 137] blue B B, 1-1/10 lb. Diamine azo blue R R, 2 lb. soda and 20 lb. Glauber's salt. Dye at the boil for one hour, rinse slightly in cold water, then enter into a fresh cold bath prepared with 4 lb nitrite of soda previously dissolved in water, and 12½ lb. hydrochloric acid. For subsequent lots in the same bath one-third of these additions is sufficient. After diazotising rinse the goods in a bath weakly acidulated with hydrochloric or sulphuric acid, and then immediately develop with beta-naphthol.

Black. — Prepare the dye-bath with 3 lb. Triamine black B, 15 lb. Glauber's salt, in fifty gallons of water. Dye exactly as in the preceding recipe. Wash and rinse very thoroughly after lifting, then diazotise in a bath of about 250 gallons of cold water, to which add separately 2½ lb. sodium nitrite dissolved in five times its bulk of water and 8 lb. hydrochloric acid diluted. Enter the damp cotton and treat it for about half an hour. Lift, pass through a weak acid bath, rinse, and develop immediately in a bath of about 250 gallons of cold water, containing 1 lb. developer T, 1 lb. soda, previously dissolved together in hot water. Enter the damp goods, work well for half an hour, then lift, wash and dry.

Blue Black.—Dye with 4 lb. Diamine black B H, 2 lb. soda and 10 lb. Glauber's salt, then diazotise and develop with naphthylamine ether.

Dark Navy.—Dye with 3 lb. Diamine azo blue R R, 2 lb. soda and 10 lb. Glauber's salt, then diazotise and develop with beta-naphthol.

Light Chestnut Brown.—Dye with 2 lb. Cotton brown N, 1 lb. diamine fast yellow A, 1 lb. soda and 10 lb. salt, then diazotise and develop with phenylene diamine.

Dark Brown.—Dye with 5 lb. Diamine cutch, 3 lb. soda and 20 lb. Glauber's salt, then diazotise and develop with fast blue developer A D.

Black.—Dye with 4 lb. Diamine black B H, 3 lb. soda and [Pg 138] 20 lb. Glauber's salt, diazotise and develop with 2 lb. resorcine and 1 lb. phenylene diamine.

Blue Black.—Dye with 4 lb. Diaminogene B, 2 lb. soda and 20 lb. Glauber's salt, then diazotise and develop with beta-naphthol.

Black.—Dye with 4½ lb. Diaminogene B, ½ oz. Diamine fast yellow B, 3 lb. soda and 20 lb. Glauber's salt, then diazotise and develop with 3 lb. resorcine and 1 lb. phenylene diamine.

Light Blue.—Dye with 1½ lb. Diaminogene blue B B, 1 lb. soda and 10 lb. Glauber's salt, then diazotised and develop with beta-naphthol.

Maroon.—Dye with 6 lb. Primuline and 20 lb. salt, diazotise and develop with blue developer A N.

Olive Brown.—Dye with 5½ lb. Diamine cutch, 3 lb. soda and 10 lb. Glauber's salt, then diazotise and develop with fast blue developer A D.

Gold Brown.—Dye with 1 lb. Cotton brown N, ¾ lb. Diamine bronze G, 2 lb. soda and 10 lb. Glauber's salt, then diazotise and develop with phenylene diamine.

Walnut Brown.—Dye with 3 lb. Diamine brown M, 3 lb. soda and 20 lb. Glauber's salt, then diazotise and develop with beta-naphthol.

Brown.—Dye with 1½ lb. Diamine brown M, 1 lb. Diamine fast yellow B, 1 lb. cotton brown N, 1 lb. soda and 10 lb. Glauber's salt, then diazotise and develop with phenylene diamine.

Dark Plum.—Dye with 3 lb. Diamine brown V, 1 lb. soda and 10 lb. Glauber's salt, then diazotise and develop with beta-naphthol.

Black Brown.—Dye with 3 lb. Diamine cutch, 3 lb. Diamine black B H, 8 lb. soda and 20 lb. Glauber's salt, then diazotised and develop with phenylene diamine.

Blue Black.—Dye with 4½ lb. Diamine black R O, 3 lb. soda and 20 lb. Glauber's salt, then diazotise and develop with beta-naphthol. [Pg 139]

Blue Black.—Dye with 4½ lb. Diamine black R O, 3 lb. soda and 20 lb. Glauber's salt, then diazotise and develop with naphthylamine ether.

Blue Black.—Dye with 5 lb. Diamine black B O, 3 lb. soda and 20 lb. Glauber's salt, then diazotise and develop with beta-naphthol.

Dark Blue.—Dye with 4 lb. Diamine black R O, 3 lb. soda and 20 lb. Glauber's salt, then diazotise and develop with blue developer A N.

Black.—Dye with 5 lb. Diamine black R O, 1 oz. Diamine bronze G, 3 lb. soda and 20 lb. Glauber's salt, then diazotise and develop with phenylene diamine.

The Diamine blacks are a range of very useful dye-stuffs, and by their means alone and in conjunction with the various developers as seen in the examples given above a range of useful shades of blue, navy blue, and blacks of every tone can be obtained. It may also be added that many of the direct dyes, although not diazotisable, are not altered by the process and so may be used along with diazotisable dyes for the purpose of shading them, and in that way a great range of shades can be produced, particularly by combining Primuline with other dyes.

(4) DIRECT DYEING FOLLOWED BY FIXATION WITH COUPLERS.

A further development in the application of the direct dyes has of late years been made. This is a two-bath method. The cotton is dyed with certain of the direct dyes: Primuline, Diamine jet blacks, Diazo blacks, Toluylene orange and brown, Diazo brown, Diamine nitrazol dyes, Benzo nitrol dyes, etc., in the usual way. Then a bath is prepared by diazotising paranitroaniline, benzidine, metanitraniline, dianisidine, etc., or by using the ready diazotised preparations which are now on the market, Nitrazol C, Azophor red P N, Azophor blue [Pg 140] P N, etc., and immersing the dyed cotton in this bath. Combination takes place between the dye on the fibre and the diazo compound in this bath, and a new product is produced direct on the fibre, which being insoluble is very resistant to washing and soaping. These "coupled" shades, as they will probably come to be called, differ from those produced on the fibre by the original dye-stuff, thus the Diamine jet blacks and some of the Diazo blacks give, with paranitroaniline, browns of various shades.

In this section also may be considered the method of dyeing cotton by using the direct colours in the ordinary way, and then "topping," as it is called, with a basic dye in a fresh bath.

Practically in the "coupling process" of dyeing only diazotised paranitroaniline is used as the coupler, although other amido bases of a similar nature are available.

When paranitroaniline is used as the source for the coupling bath it is well to prepare a stock bath of diazotised paranitroaniline, which may be done in the following manner:—

Preparation of diazotised paranitroaniline.—Take 1 lb. paranitroaniline, mix with 1 gallon boiling water and 1 quart hydrochloric acid, stir well, when the paranitroaniline will dissolve the solution may if necessary be assisted by a little heat. Now add 1½ gallons of cold water, and set aside to cool, when the hydrochloride of paranitroaniline will separate out in the form of fine crystals; when the mixture is quite cold (it cannot be too cold) there is added ½ lb.

sodium nitrite dissolved in ½ gallon cold water, stir well for fifteen to twenty minutes, by the end of which time the paranitroaniline will have become fully diazotised, cold water is added to bring up the volume of the mixture to 10 gallons. This stock bath well prepared and kept in a cool, dark place will keep good for three to four weeks. This bath contains 1 lb. of paranitroaniline in 10 gallons, and it is a good rule [Pg 141] to allow ½ lb., or 5 gallons of this stock bath to each pound of dye-stuff used in dyeing the ground colour to be developed up.

To prepare the coupling bath there is taken 5 gallons of the stock bath, 1 lb. sodium acetate with sufficient water for each 1 lb. of dye that has been used.

This bath is used cold, and the cotton is worked in it for half an hour, then it is taken out, washed well and dried.

Nitrazol C is a ready prepared diazotised paranitroaniline in a powder form which keeps well if stored in a dry place. The method of using is to take 8 lb. Nitrazol C, stir into a paste with water and then add this paste to the coupling bath, together with 2 lb. soda and ¾ lb. acetate of soda. This bath is used cold and the dyed cotton is immersed in it for half an hour, then taken out, well washed and dried.

The quantity of Nitrazol C given will suffice for all shades dyed with from 2 to 4 per cent, of dye-stuff, but when paler shades are dyed, using less than say ½ per cent. of dye-stuff, about 4 lb. Nitrazol C, with the soda and acetate of soda in proportionate quantities, may be used.

Azophor red P N is also a preparation of diazotised paranitroaniline in the form of a dry powder which keeps well.

To prepare the coupling bath there is taken 2 lb. of Azophor red P N, which is dissolved in water and added to the bath along with 1 lb. acetate of soda. The dyed goods are worked in the cold bath for half an hour, then taken out, well washed and dried.

The quantities given are sufficient for shades dyed with 2 to 4 per cent. of dye-stuff; for weaker shades half the quantities may be taken.

Benzo-nitrol developer is sold in the form of a yellow paste. To use it take 5 lb., stir into a smooth paste with water, then add to the coupling bath. There is then added [Pg 142] 3 pints of hydrochloric acid, with some stirring. Allow to stand for half an hour, add 1½ lb. acetate of soda and 6½ oz. soda, when the bath is ready for use. The cotton is entered and worked for half an hour, then lifted out, washed and dried.

It may be mentioned that solutions of the three couplers just named may be kept for some time without decomposition, but as soon as soda and acetate of soda are added they begin to decompose and then cannot be kept more than a few hours in a good condition. It is a good plan therefore not to add the acetate of soda until the bath is to be used.

An excess of coupler in the bath does no harm, but a deficiency may lead to poor and weak shades being developed.

The following recipes show the dyes which may be applied by this method and give some idea of the colours that can be got. Only the dye-stuffs are given. Any of the above couplers can be used with them as may be most convenient.

Black.—Dye with 5 lb. Benzo-nitrol black B, 1 lb. soda and 20 lb. Glauber's salt.

Olive Green.—Dye with 6 lb. Primuline, 3 lb. Titan ingrain blue and 20 lb. salt.

Black.—Dye with 4 lb. Dianil black C R, 2 lb. soda and 25 lb. salt.

Dark Blue.—Dye with 2 lb. Dianil dark blue R, 1 lb. Dianil dark blue 3 R, 2 lb. soda and 25 lb. salt.

Gold Brown.—Dye with 1 lb. Primuline, 8 oz. Dianil brown R and 20 lb. salt.

Chestnut.—Dye with 3 lb. Primuline, ¾ lb. Dianil brown G O, 1 lb. Dianil brown E, 1 lb. soda and 20 lb. salt.

Dark Brown.—Dye with 1 lb. Dianil brown 3 G O, 3 lb. Dianil brown D, 1 lb. soda and 20 lb. salt.

Dark Green.—Dye with 4 lb. Primuline, 1½ lb. Dianil black C R, 1 lb. soda and 20 lb. salt.

Walnut Brown.—Dye with 1 lb. Dianil brown 3 G O, 8 oz. Dianil brown R, 3 lb. Dianil brown B D, 1 lb. soda, and 20 lb. salt. [Pg 143]

Light Green.—Dye with 3 lb. Primuline, 8 oz. Dianil blue B, 5 oz. Dianil dark blue R, 1 lb. soda, and 20 lb. salt.

Orange Yellow.—Dye with 3¼ lb. Primuline, 1 lb. Oxydianil yellow, and 25 lb. salt.

Olive.—Dye with 3½ lb. Primuline, 8 oz. Dianil brown 3 G O, 8 oz. Dianil blue B, 4 oz. Dianil dark blue R, 1 lb. soda, and 25 lb. salt.

Bright Yellow.—Dye with 2 lb. Primuline, and 20 lb. salt.

Gold Yellow.—Dye with 2 lb. Diamine fast yellow A, 1 lb. soda, and 20 lb. salt.

Bright Walnut.—Dye with ½ lb. Diamine nitrazol brown B, 1 lb. Oxydiamine orange R, 1 lb. soda, and 20 lb. Glauber's salt.

Gold Brown.—Dye with ½ lb. Diamine nitrazol brown G, 1 lb. Primuline, and 20 lb. salt.

Green.—Dye with 2 lb. Primuline, 1 lb. Diamine nitrazol black B, 1 lb. soda and 20 lb. salt.

Pale Chestnut.—Dye with 1 lb. Primuline, ½ lb. Oxydiamine orange R and 20 lb. salt.

Moss Brown,—Dye with 2 lb. Primuline, 1 lb. diamine jet black O O and 20 lb. salt.

Chocolate.—Dye with 1½ lb. Diamine brown V, 2 lb. Diamine nitrazol brown R D, 2 lb. soda and 20 lb. Glauber's salt.

Olive Brown.—Dye with 2 lb. Diamine nitrazol brown G, 1 lb. Diamine nitrazol black B, 1 lb. soda and 20 lb. Glauber's salt.

Russian Green.—Dye with 2 lb. Diaminogene extra, 2 lb. soda and 20 lb. Glauber's salt.

Bronze Green.—Dye with 2 lb. Diamine grey G, 2 lb. soda and 20 lb. Glauber's salt.

Terra-cotta Bed.—Dye with 2 lb. Oxydiamine orange R, 1 lb. soda and 20 lb. Glauber's salt.

Terra-cotta Brown.—Dye with 2 lb. Diamine nitrazol brown R D, 1 lb. soda and 20 lb. Glauber's salt. [Pg 144]

Olive Green.—Dye with 1 lb. Primuline, 2 lb. Diamine bronze G, 1 lb. soda and 20 lb. Glauber's salt.

Dark Green.—Dye with 1 lb. Primuline, 2 lb. Diamine nitrazol black B, 2 lb. soda and 20 lb. salt.

Sage Brown.—Dye with 1 lb. Primuline, 2 lb. Diamine jet black O O, 1 lb. soda and 20 lb. salt.

Black Brown.—Dye with 1 lb. Diamine brown V, 2 lb. Diamine nitrazol black B, 2 lb. soda and 20 lb. Glauber's salt.

Dark Walnut.—Dye with 1 lb. Diamine brown V, 2 lb. Oxydiamine orange R, 2 lb. soda and 20 lb. Glauber's salt.

Pale Sage.—Dye with 1 lb. Diamine brown V, 2 lb. Primuline, 1 lb. soda and 20 lb. salt.

Brown.—Prepare the dye-bath with 3 lb. Diamine jet black O O, 20 lb. Glauber's salt, 2 lb. soda. Dye at the boil for one hour.

Brown.—Prepare the dye-bath with ¾ lb. Benzo nitrol brown G, 20 lb. Glauber's salt, 2 lb. soda. Dye for one hour at the boil.

Dark Brown.—Prepare the dye-bath with 2 lb. Benzo nitrol dark brown N, 20 lb. Glauber's salt, 2 lb. soda. Dye for one hour at the boil.

Brown.—Prepare the dye-bath with 4 lb. Direct fast brown B, 20 lb. Glauber's salt, 2 lb. soda. Dye for one hour at the boil.

Brown.—Prepare the dye-bath with 1 lb. 11 oz. Diamine jet black O O, 2 lb. Cotton brown N, 1 lb. 5 oz. Diamine brown V, 20 lb. Glauber's salt, 2 lb. soda. Dye at the boil for one hour.

Brown.—Prepare the dye-bath with 2 lb. Diamine bronze G, 6½ oz. Cotton brown N, 9¾ oz. Diamine fast yellow A, 20 lb. Glauber's salt, 2 lb. soda.

Black.—Prepare the dye-bath with 5 lb. Pluto black B, 20 lb. Glauber's salt, 2 lb. soda. Dye for one hour at the boil. [Pg 145]

Solidogen A is a new coupler that has latterly been applied. It is a syrupy liquid, and the coupling bath is made by taking from 4 lb. to

6 lb. of the Solidogen A, and 1 lb. to 2 lb. of hydrochloric acid, in place of which 3 lb. to 5 lb. alum may be used. This bath is used at the boil, the goods being treated for half an hour, then well rinsed and dried. It increases the fastness of the colours to washing and soaping.

The following recipes show its application: —

Bright Bed. — Dye with 3 lb. Dianil red 4 B, 2 lb. soap, 3 lb. soda and 15 lb. Glauber's salt, then fix with Solidogen A.

Scarlet. — Dye with 3 lb. Dianil scarlet G, 2 lb. soda and 25 lb. salt; fix with Solidogen A.

Plum. — Dye with 3 lb. Dianil claret B, 5 lb. soda and 10 lb. Glauber's salt, then fix with Solidogen A.

Topping with Basic Dyes. — The shades dyed with the direct dyes may be materially brightened and new shades produced by topping with any of the basic dyes, which are applied in a fresh warm bath. A great variety of effects may be thus got of which the following recipes give a few examples: —

Green. — Dye with 1 lb. Titan yellow G and 20 lb. salt; top with ½ lb. Brilliant green.

Blue. — Dye with 1¾ lb. Diamine azo blue R, 1 lb. soda and 20 lb. Glauber's salt, then top with 2 oz. New Methylene blue N.

Bright Blue. — Dye with ¾ lb. Diamine brilliant blue G, 1 lb. soda and 10 lb. Glauber's salt; top with 2 oz. New Methylene blue 3 R.

Blue. — Dye with 1 lb. Diamine sky blue, 1 lb. soda and 10 lb. Glauber's salt, and top with 4 oz. Brilliant green.

Bose Lilac. — Dye with 1½ oz. Diamine violet N, 1 lb. soda and 10 lb. Glauber's salt, then top with 2 oz. Tannin heliotrope.

Green. — Dye at the boil for one hour with 2 lb. Benzo [Pg 146] green G and 10 lb. Glauber's salt, then top in a fresh bath with ½ lb. Turquoise blue B B.

Violet. — Dye with 5 oz. Diamine violet N, 2 oz. Diamine brilliant blue G, 1 lb. soda and 10 lb. salt, and top with 1 oz. Methyl violet 2 B.

Plum.—Dye with 1½ lb. Oxydiamine violet B, 5 oz. Diamine red 10 B, 2 lb. soda and 10 lb. Glauber's salt, then top with 1½ oz. Methyl violet R.

Bright Green.—Dye with 1¼ lb. Diamine green G, 1¼ lb. Oxydiamine yellow G G, 2 lb. soda and 10 lb. Glauber's salt, then top with 2 oz. Brilliant green.

Blue.—Dye with 2 lb. Benzo azurine G, 3 oz. Brilliant azurine B, 1 lb. soda and 20 lb. Glauber's salt, topping with 6 oz. Turquoise blue G and 3 oz. New Victoria blue B.

Dark Lilac.—Dye with 3¾ lb. Heliotrope B B, 1 lb. soda and 20 lb. Glauber's salt, then top with 1 lb. Methyl violet R, and ½ lb. Methyl violet 3 R.

Scarlet.—Dye with 3 lb. Brilliant Congo R, 3 lb. soda and 20 lb. Glauber's salt, then top with 8 oz. Safranine.

Bright Green.—Dye with 3 lb. Chrysamine G, 2 lb. soap and 10 lb. phosphate of soda, topping with ¾ lb. Malachite green.

Bright Violet.—Dye with 1½ lb. Chicago blue 6 B, 1 lb. soda and 20 lb. Glauber's salt, topping with 10 oz. Methyl violet B.

Dark Green.—Dye with 2 lb. Columbia green, 3 lb. soda and 10 lb. Glauber's salt, topping with 10 oz. Malachite green.

Claret.—Prepare a dye-bath with ¾ oz. Diamine black R O, 2½ lb. Benzo purpurine 6 B, 10 lb. Glauber's salt. Dye at the boil for one hour, then enter in a fresh cold bath of ½ lb. Safranine G. Work for twenty minutes, lift, wash and dry.

Seal Brown.—Make up a dye-bath with 2 lb. Benzo azurine G, 20 lb. Glauber's salt. Enter yarn at 180° F., dye at the boil for one hour, lift, wring, and enter into a fresh bath of [Pg 147] 1½ lb. Bismarck brown. Work for one hour at about 180° F., lift, rinse well and dry.

(5) DYEING ON TANNIC MORDANT.

The oldest group of coal-tar dyes are the basic dyes, of which Magenta, Brilliant green, Chrysoidine, Bismarck brown, Auramine are typical representatives. For a long time these dyes were only used for dyeing wool and silk; for cotton, linen, and some other vegetable fibres they have little or no affinity, and hence cannot dye them direct. However, it was found out that if the cotton be prepared or mordanted (as it is called) with tannic acid or with any substance containing that compound they could be used for dyeing cotton.

The mordant used, tannic acid, has the property of combining with the dyes of this group to form insoluble coloured tannates. Now tannic acid has a certain amount of affinity for cotton, if the latter be immersed in solution of tannic acid or any material containing it some of the latter is taken up and more or less fixed by the cotton fibre. Tannic acid is a vegetable product found in a large number of plants, and plant products, such as sumac, myrabolams, divi-divi, galls, oak bark, gambier, cutch, algarobilla, valonia, etc., which are commonly known as tannins, or tannin matters, on account of their use in the conversion of animal skins or hides into leather, which is done in the tanning industry.

By itself the tannin-colour lake, which may be formed on the cotton fibre by immersion first in a bath of tannin and then in a dye-bath, is not fast to washing and soaping, but by taking advantage of the fact with such metals as tin, iron, antimony, etc., it combines to form insoluble tannates; the tannic acid can be fixed on the cotton by immersion in a bath containing such fixing salts as tartar emetic, tin crystals, copperas, antimony fluoride, and antimony oxalate. The dye [Pg 148] ing of cotton with the basic colours therefore resolves itself into three operations: —

(1) Tanning with tannic acid or some tanning matters.

(2) Fixation with tartar emetic or other fixing agent.

(3) Dyeing with the required colour or mixture of colours.

(1) **The Tanning Operation.** — The practice of tanning or mordanting cotton with tannin is variously carried on by dyers. Some steep the cotton in the tannin bath over night, others immerse it from two to three hours in a lukewarm bath, while some enter it in a boiling bath, which is then allowed to cool down and the cotton is lifted out. The last is perhaps the quickest method, and experiments have shown that it is as good as any other method, if the quantity of tannic acid taken up be regarded as the criterion of success.

In the natural products which have just been enumerated, the tannic acid is accompanied by some colouring matter, which is also absorbed by the cotton; in some (sumac and galls) this colour is present in but small quantities; in others (divi-divi, myrabolams, algarobilla), there is a large quantity; therefore cotton treated with these comes out more or less coloured. Now it is obvious that such forms of tannin cannot be used when light tints are to be dyed, for such the acid itself must be used, for medium shades sumac or galls may be used; while when dark shades—browns, maroons, dark greens, navy blues, etc., are to be dyed, then such tannin matters as divi-divi or myrabolams may be conveniently and economically adopted. The quantity used varies according to circumstances; the character of the shades that are to be dyed, the dye-stuff used, the quality and character of the tannin matter used. For pale shades about 1 per cent. of tannic acid may be used, deep shades require from 3 to 4 per cent. Of tannin matters from 5 per cent. may be used for pale shades, from 20 to 25 per cent. for deep shades. The [Pg 149] tannin baths are not exhausted, and may be kept standing, adding for each succeeding lot of cotton from ½ to ¾ of the above quantities of tannin matters. Of course sooner or later the baths become unusable from various causes, and then they may be thrown away; but old tannin baths often work better than the new ones.

(2) **The Fixing Bath.**—Following on the tannin bath comes the fixing bath to fix the tannin on the cotton in the form of insoluble metallic tannates. A variety of metallic salts may be used for this purpose, those of antimony, tin, iron, lead, etc., the substances most commonly used being tartar emetic, antimony fluoride, oxalate of antimony, tin crystals and copperas.

Beyond forming the insoluble tannate of antimony or tin, as the case may be, the salts of antimony and tin have no further effect on the tanned cotton, and they may be used to fix the tannin for all tints or shades, from very pale to very deep. Of all these salts tartar emetic has been found to be the best, probably because it is the least acid in its reactions, and therefore there is no tendency to remove any tannin from the fibre, as is the case with the other salts. Tin salt is little used for this purpose, because of its acidity, which prevents it from fixing the tannin as completely as is the case with tartar emetic.

With copperas or iron liquor the question comes up not only of the mere fixation of the tannin, but also the fact that iron forms with tannin grey to black compounds, hence cotton which has been tanned and then immersed in a bath of copperas becomes coloured grey to black, according to the quantity of the tannin matter used. The property is useful when dark shades of maroons, clarets, greens, browns, blues, etc., are to be dyed, and is frequently employed.

(3) **Dyeing.**—After the tannin and fixing operations comes the dyeing. This is not by any means a difficult operation. [Pg 150] It is best carried out by preparing a cold bath, entering the prepared or mordanted cotton in this and heating slowly up. It is not necessary to raise to the boil nor to maintain the dye-bath at that heat, a temperature of 180° F. being quite sufficient in dyeing with the basic colours, and the operation should last only until the colouring matter is extracted from the dye-bath. Working in this way, level uniform shades can be got.

One source of trouble in the dyeing of basic dyes, more especially with magentas, violets and greens, lies in their slight solubility and great strength. In preparing solutions of such dyes it is best to dissolve the dye-stuff by pouring boiling water over it, and stirring well until all appears to be dissolved.

This solution should be strained through a fine cloth, as any undissolved specks will be sure to fix themselves on the cloth and lead to dark spots and stains, as, owing to the weak solubility of the dye, and this being also fixed as insoluble tannate by the tannic acid on the fibre, there is no tendency for the dye to diffuse itself over the cloth, as occasionally happens in other methods of dyeing. No advantage is gained by adding to the dye-bath such substances as common salt or Glauber's salt.

Some few of the dyes, *e.g.*, Soluble blue, Victoria blue, which will dye on a tannic mordant, are sulphonated compounds of the colour base. These can be dyed in medium or light shades on to unmordanted cotton from a bath containing alum, and in the ordinary method of dyeing the addition of alum is beneficial, as tending to result in the production of deep shades. These are somewhat faster to washing and soaping, which is owing to the dye-stuff combining with the alumina in the alum to form an insoluble colour lake of the sulphuric acid groups it contains.

Many recipes can be given for the dyeing of basic dyes on cotton; for the range of tints and shades of all colours [Pg 151] that can be produced by their means is truly great. Typical recipes will be given showing what basic dyes are available and how they can be combined together. The dyer, however, who knows how to produce shades by combining one dye-stuff with another is able to get many more shades than space will permit to be given here.

Sky Blue.—Mordant with 1 lb. tannic acid and ½ lb. tartar emetic, then dye with 2½ oz. Victoria blue B, and ½ oz. Turquoise blue G.

Bright Blue.—Mordant with 2 lb. tannic acid and 1½ lb. tartar emetic, then dye with 1½ lb. New Victoria blue B, and ¾ lb. Turquoise blue G.

Pale Green.—Mordant with 1½ lb. tannic acid and ¼ lb. tartar emetic, then dye with ¼ oz. Brilliant green and ½ oz. Auramine I I.

Bright Green.—Mordant with 1 lb. tannic acid and 1 lb. tartar emetic, then dye with 6 oz. Brilliant green and 8 oz. Auramine I I.

Turquoise Blue.—Mordant with 2 lb. tannic acid and 1½ lb. tartar emetic; dye with 1 lb. Turquoise blue G.

Crimson.—Mordant with 3 lb. tannic acid and 2 lb. tartar emetic, then dye with 1¼ lb. Brilliant rhoduline red B and 5 oz. Auramine I I.

Bright Violet.—Mordant with 2 lb. tannic acid and 1 lb. tartar emetic; dye with 1¼ lb. Rhoduline violet and 3 oz. Methyl violet B.

Rose Lilac.—Mordant with 2 lb. tannic acid and 1 lb. tartar emetic; dye with ¾ lb. Rhoduline violet.

Yellow.—Mordant with 3 lb. tannic acid and 2½ lb. tartar emetic; dye with 2 lb. Auramine I I.

Orange.—Mordant with 3 lb. tannic acid and 2¼ lb. tartar emetic; dye with 2 lb. Chrysoidine.

Green.—Mordant with 3 lb. tannic acid and 2½ lb. tartar emetic, then dye with 2 lb. Brilliant green. [Pg 152]

Red Violet.—Mordant with 1½ lb. tannic acid and 1¼ lb. tartar emetic, then dye with 8 oz. Methyl violet 4 R.

The same method may be followed with all the brands of Methyl, Paris and Hoffmann violets, and so a range of shades from a purple to a pure violet can be dyed. The 4 R to R brands of these violets dye reddish shades, the redness decreasing according to the mark, the B to 6 B brands dye bluish shades, the blueness increasing till the 6 B brand dyes a very blue shade of violet.

Bright Sea-Green.—Mordant with 1 lb. tannic acid and ½ lb. tartar emetic, then dye with 2 oz. Turquoise blue G and ¼ oz. New Victoria blue B. With these two blues a very great variety of tints and shades of blue can be dyed.

Sea Green.—Mordant with 1 lb. tannic acid and ¾ lb. tartar emetic, and dye with 2 oz. Malachite green.

Dark Green.—Mordant with 3 lb. tannic acid and 2 lb. tartar emetic, then dye with 1¼ lb. Malachite green and ¾ oz. Auramine I I.

By combination of Brilliant green or Malachite green (which are the principal basic greens) with Auramine in various proportions a great range of greens of all tints and shades, from pale to deep and from very yellow to very green tints, can be dyed.

Scarlet.—Mordant with 3 lb. tannic acid and 2 lb. tartar emetic, then dye with 1½ lb. Rhodamine 6 G and ¼ lb. Auramine I I.

Pink.—Mordant with 1 lb. tannic acid and ¾ lb. tartar emetic, and dye with ¼ lb. Rhodamine G.

Pink.—Mordant with 1 lb. tannic acid and ¾ lb. tartar emetic, and dye with ½ lb. Irisamine G.

Slate Blue.—Mordant with 1½ lb. tannic acid and 1 lb. tartar emetic, then dye with ¾ lb. Victoria blue B and 2 oz. Malachite green.

Indigo Blue.—Mordant with 3 lb. tannic acid and 2½ lb. [Pg 153] tartar emetic, then dye with 1 lb. Methylene blue 2 B and ½ lb. Malachite green.

Dark Scarlet.—Mordant with 3 lb. tannic acid and 2½ lb. tartar emetic, then dye with 1 lb. Safranine prima and ½ lb. Auramine I I.

Grey.—Mordant with 1 lb. tannic acid and ½ lb. tartar emetic, then dye with ½ lb. New Methylene grey B.

Bluish Rose.—Mordant with 2 lb. tannic acid and 1 lb. tartar emetic, then dye with 1 lb. Rhodamine B.

Maroon.—Mordant with 4 lb. tannic acid and 2 lb. tartar emetic, and dye with 2 lb. Magenta and ½ lb. Auramine.

Dark Green.—Mordant with 5 lb. tannic acid and 2½ lb. tartar emetic, and dye with ½ lb. Methylene blue B B, ½ lb. Methyl violet 2 B and 2½ lb. Auramine I I.

Orange.—Mordant with 3 lb. tannic acid and 2 lb. tartar emetic, and dye with 1 lb. New Phosphine G.

Lilac Grey.—Mordant with 1 lb. tannic acid and 1½ lb. tartar emetic, and dye with ¼ lb. Methylene grey B F.

Gold Brown.—Mordant with 3 lb. tannic acid and 1½ lb. tartar emetic, and dye with ¾ lb. Thioflavine T and ¾ lb. Bismarck brown.

Orange.—Mordant with 2 lb. tannic acid and 1 lb. tartar emetic, and dye with 1 lb. Auramine and ¼ lb. Safranine.

Dark Blue.—Mordant with 3 lb. tannic acid and 2 lb. tartar emetic, and dye with 2¼ lb. New Methylene blue R and ¾ lb. Naphtindone B B.

Olive Green.—Mordant with 5 lb. sumac extract and 2 lb. copperas, and dye with 1 lb. Auramine.

Russian Green.—Mordant with 5 lb. sumac extract and 2 lb. copperas, and dye with 2 lb. Malachite green.

Scarlet.—Mordant with 3 lb. tannic acid and 2 lb. tartar emetic, and dye with 1½ lb. Thioflavine T and ½ lb. Irisamine G. With these two dye-stuffs it is possible to produce a variety of useful shades from a pure greenish yellow, with [Pg 154] Thioflavine T alone, to a bright bluish pink, with the Irisamine alone, through orange, scarlet, etc., with combinations of the two dye-stuffs.

Dark Grey.—Mordant with 5 lb. sumac extract and 3 lb. copperas, then dye with 1 lb. New Methylene grey G.

Blue Black.—Mordant with 8 lb. sumac extract and 4 lb. copperas, or better with iron liquor, then dye with 2 lb. Indamine blue N.

Olive Brown.—Mordant with 5 lb. sumac extract and 3 lb. copperas, and dye with 1½ lb. New Phosphine G.

Indigo Blue.—Mordant with 1½ lb. tannic acid and 1 lb. tartar emetic; dye with ½ lb. New Methylene blue N.

Sky Blue.—Mordant with ½ lb. tannic acid and ¾ lb. tartar emetic; dye with 1½ oz. New Methylene blue G G.

Dark Violet.—Mordant with 3 lb. tannic acid and 2 lb. tartar emetic, then dye with 4 lb. Fast neutral violet B.

Bright Yellow.—Mordant with 2 lb. tannic acid and 1 lb. tartar emetic, and dye with 2 lb. Thioflavine T.

Primrose Yellow.—Mordant with 1 lb. tannic acid and ¼ lb. tartar emetic, and dye with 2 oz. Thioflavine T.

Navy Blue.—Mordant with 5 lb. sumac extract and 3 lb. copperas, then dye with 2 lb. New Methylene blue R.

Violet.—Mordant with 3 lb. tannic acid and 2 lb. tartar emetic, and dye with 2 lb. New Methylene blue 3 R.

Dark Blue.—Mordant with 5 lb. sumac extract and 3 lb copperas, and dye with 2 lb. New Methylene blue N X.

Blue Black.—Mordant with 8 lb. sumac extract and iron liquor, then dye with 3 lb. Metaphenylene blue B.

Emerald Tint.—Mordant the cotton in the usual way with 1 lb. tannic acid and 1 lb. tartar emetic; dye to shade at 180° F. in a bath containing 14 oz. Auramine G, 2 oz. Brilliant green, then lift, wash and dry.

Orange.—Mordant with 3 lb. tannic acid and 2 lb. tartar emetic, then dye with 4 lb. Tannin orange R. [Pg 155]

Scarlet.—Mordant with 3 lb. tannic acid and 2 lb. tartar emetic, and dye with 2 lb. Tannin orange R and 1 lb. Safranine S.

Dark Scarlet.—Mordant with 3 lb. tannic acid and 2 lb. tartar emetic, and dye with ½ lb. Tannin orange R and 2 lb. Safranine S.

The Janus colours are a series of dyes of a basic nature which can be applied somewhat differently to the ordinary basic dyes, although the ordinary method can be followed. With these Janus dyes a two-bath process is followed. A dye-bath is prepared containing the dye-stuff, sulphuric acid and common salt, and this is used at the boil from half to three-quarters of an hour, and the goods are allowed to remain in another three-quarters of an hour while the bath cools down. Next the dyed goods are run in a fixing bath of sulphuric acid, tannic acid and tartar emetic, this is used at the boil from half to one hour, after which the dyed goods are taken out and washed. If necessary the goods may be now topped with basic colours in order to produce any desired shade. The following recipes will show how the Janus dyes may be used: —

Blue. — Dye with 9 oz. sulphuric acid, 8 oz. Janus blue G, and 5 lb. common salt; fix with ¾ lb. sulphuric acid, 8 oz. tannic acid and 4 oz. tartar emetic.

Turquoise Blue. — Dye with 9 oz. sulphuric acid, 1 lb. Janus green B and 10 lb. salt, fixing with ¾ lb. sulphuric acid, 1 lb. tannin and ½ lb. tartar emetic.

Dark Blue. — Dye with 9 oz. sulphuric acid, 2½ lb. Janus blue R and 15 lb. common salt; fix with ¾ lb. sulphuric acid, 2½ lb. tannic acid and 1¼ lb. tartar emetic.

Buff. — Dye with 9 oz. sulphuric acid, 2 oz. Janus yellow R and 3 lb. salt; fix with ¾ lb. sulphuric acid, 3 oz. tannic acid and 3 lb. tartar emetic.

Crimson. — Dye with 9 oz. sulphuric acid, 2½ lb. Janus red [Pg 156] B and 15 lb. salt, fixing with ¾ lb. sulphuric acid, 2½ lb. tannic acid and 1¼ lb. tartar emetic.

Red Violet. — Dye with 9 oz. sulphuric acid, 1 lb. Janus claret red B and 10 lb. salt; fix with 12 oz. sulphuric acid, 1 lb. tannic acid and ½ lb. tartar emetic.

Orange. — Dye with 9 oz. sulphuric acid, 1 lb. Janus yellow R and 10 lb. salt; fix with 12 oz. sulphuric acid, 1 lb. tannic acid and ½ lb. tartar emetic.

Dark Violet.—Dye with 9 oz. sulphuric acid, 2 lb. Janus grey B and 15 lb. salt; fix with 12 oz. sulphuric acid, 2½ lb. tannic acid and 1¼ lb. tartar emetic.

Chocolate Brown.—Dye with 9 oz. sulphuric acid, 3½ lb. Janus brown B and 15 lb. salt, fixing with 2 oz. sulphuric acid, 2½ lb. tannic acid and 1 lb. tartar emetic.

(6) DYEING ON METALLIC MORDANTS.

There are a number of dye-stuffs or colouring matters like alizarine, logwood, fustic, barwood, cutch, resorcine green, etc., which have no affinity for the cotton fibre, and of themselves will not dye it. They have the property of combining with metallic oxides such as those of iron, chromium, aluminium, tin, lead, calcium, etc., to form coloured bodies which are more or less insoluble in water. These coloured bodies are called "colour lakes," and the metallic compounds used in connection with their production "mordants," hence often the dye-stuffs applied by this method are termed "mordant dyes". In the case of the natural dye-stuffs—logwood, fustic, Persian berries, Brazil wood, camwood, cochineal, quercitron, cutch, etc.—which belong to this group of "mordant dyes," the whole of the material does not enter into the operation, but only a certain constituent contained therein, which is commonly soluble in boiling water, and extracted out by boiling. This constituent is called the "colouring principle" of the dye-stuff or wood, and naturally varies with each. It is not [Pg 157] intended here to deal in detail with these colouring principles. The methods of applying and the colours which can be got from these dyes varies very much. Roughly, the modes of application fall under three heads: (1) the particular metallic mordant is first fixed on the fibre by any suitable method, and then the fibre is dyed; (2) the dye-stuff is first applied to the fibre, and then the colour is fixed and developed by treatment with the mordant; and (3) the dye-stuff and the mordant are applied at the same time. This last method is not much used. In the following sections many examples of these methods will be given.

The dyes fixed with metallic mordant vary in their composition and properties. There is first the group of eosine dyes, which are acid derivatives of a colour-base, and, in virtue of being so, will combine with the metallic oxides. The colour of these colour lakes is quite independent of what oxide is used, depending only on that of the particular eosine dye employed. Then there are some members of the azo dyes, particularly the croceine scarlets, which can also be dyed on the cotton by the aid of tin, lead or alum mordants. Here, again, the mordant has no influence on the colour, but only fixes it on the cotton.

The most important class of dye-stuffs which are dyed on to cotton with a metallic mordant is that to which the term "mordant dyes" is now given. This includes such dyes as logwood, fustic, madder, alizarine, and all the dyes derived from anthracene. Many of these are not really dyes, that is, they will not of themselves produce or develop a colour on to any fibre when used alone; it is only when they combine with the mordant oxide which is used, and then the colour varies with the mordant. Thus, for instance, logwood with iron produces a bluish black; with chrome, a blue; with alumina, a reddish blue. Alizarine with iron produces a dark violet; with alumina, a scarlet; with chrome, a red; with [Pg 158] tin, a bright scarlet. Fustic gives with tin and alumina, bright yellows; with chrome, a dark yellow; with iron, an olive, and so on with other members of this group, of which more will be said later on.

Dyeing with Eosines.

At one time a fairly large quantity of cotton was dyed with the eosines, owing to the brightness of the shades given by them; but the introduction of such direct dyes as the Erikas, Ceranines, etc., has thrown the eosines out of use.

The method adopted for the production of eosine pinks and scarlets on cotton involves three operations: (1) impregnating the cloth with sodium stannate; (2) fixing oxide of tin by a bath of weak sulphuric acid; and (3) dyeing with the eosine.

(1) Preparing with Sodium Stannate.—A bath of 8° Tw. is prepared, and the cotton is allowed to steep in this bath until it becomes thoroughly impregnated, after which it is taken out and wrung.

(2) Fixing the Tin Oxide.—A bath of sulphuric acid of 2° to 4° Tw. is prepared, and the cotton is sent through it, after which it is washed well with water, when it is ready for dyeing.

Stannate of soda is easily decomposed by acids; even the carbonic acid present in the air will bring about this change. The tin contained in the stannate is deposited on the cotton in the form of stannic oxide, or, more strictly, stannic acid. As this is somewhat soluble in acids, it is important that the sulphuric acid bath be not too strong, or there will be a tendency for the tin oxide to be dissolved off the cotton, and then but weak shades will be obtained in the final operation of dyeing. Further, owing to the decomposition of the stannate by exposure to the air, it is important that the substance should be used while fresh, and that only fresh baths should be used. [Pg 159]

(3) Dyeing with Eosine Colours. — After the treatment with stannate of soda and sulphuric acid the prepared cotton is ready for dyeing. This process is carried out by preparing a cold bath with the required dye-stuff, entering the cotton therein, and then slowly raising to about 180° F., and maintaining at that heat until the desired shade is obtained. It is not needful to raise to the boil and work at that heat. No better results are obtained, while there is even a tendency for colours to be produced that rub badly, which is due to the too rapid formation of the colour lake; and it is worthy of note that when a colour lake is rapidly formed on the fibre in dyeing it is apt to be but loosely fixed, and the colour is then loose to both washing and rubbing.

Dyeing with Acid and Azo Dyes.

In dyeing with this class of colours stannate of soda, acetate of lead or alum may be used as mordants. The stannate of soda is employed in the same manner as when the eosines are used, and, therefore, does not require to be further dealt with.

Acetate of lead is used in a similar way. The cotton is first steeped in a bath of acetate of lead of about 10° Tw. strong, used cold, and from half an hour to an hour is allowed for the cotton to be thoroughly impregnated with the lead solution, it is then wrung and passed a second time into a bath of soda, when lead oxide or lead carbonate is deposited on the cotton. After this treatment the cotton is ready for dyeing with any kind of acid, azo and even eosine dyes, and this is done in the same manner as is used in dyeing the eosines on a stannate mordant. The shades obtained on a lead mordant cannot be considered as fast; they bleed on washing and rub off badly.

When alum is used as the mordant it may be employed in the same way as acetate of lead, but as a rule it is added to [Pg 160] the dye-bath direct, and the dyeing is done at the boil. This latter method gives equally good results, and is more simple.

The eosines and erythrosines, water blues, soluble blues, croceine scarlets, cloth scarlets, and a few other dyes of the azo and acid series are used according to this method. The results are by no means first class, deep shades cannot be obtained, and they are not fast to washing, soaping and rubbing.

The methods of employing the much more important group of colouring matters known as the mordant dyes, which comprise such well-known products as logwood, fustic and alizarine, require more attention. With these, alumina, iron, and chromium mordants are used as chief mordants, either alone or in combination with one another, and with other bodies. The principal point is to obtain a good deposit of the mordant on the cotton fibre, and this is by no means easy.

There are several methods by the use of which a deposit is formed of the mordant, either in the form of metallic oxide (or, perhaps, hydroxide) or of a basic salt. In some cases the cotton is passed through alternate baths containing, on one hand, the mordanting salt, *e.g.*, alum, copperas, etc., and, on the other, a fixing agent, such as soda or phosphate of soda. Or a mordanting salt may be used, containing some volatile acid that on being subjected to a subsequent steaming is decomposed. Both these methods will be briefly discussed.

Methods of Mordanting.

The cotton is first steeped in a bath containing Turkey-red oil, and is then dried. By this means there is formed on the fibre a deposit of fatty acid, which is of great value in the subsequent dyeing operations to produce bright and fast shades. After the oiling comes a bath of alum or alumina sulphate, either used as bought, or made basic by [Pg 161] the addition of soda. The result is to bring about on the fibre a combination of the fatty acid with the alumina. Following on the alum bath comes a bath containing soda or phosphate of soda, which brings about a better fixation of the alumina.

These operations may be repeated several times, especially when a full shade having a good degree of fastness is desired, as, for instance, Turkey-reds from Alizarine. This method of mordanting is subject to considerable variations as regards the order in which the various operations are carried out, the strength of the baths, and their composition. A great deal depends upon the ultimate result desired to be obtained, and the price to be paid for the work.

Iron is much easier to fix on cotton than is either alumina or chrome. It is usually sufficient to pass the cotton through a bath of either copperas or iron liquor, hang up to dry or age, and then pass into a bath of lime, soda or even phosphate of soda. The other mordants require two passages to ensure proper deposition of the mordant on the fibre.

Following on the mordanting operations comes the dyeing, which is carried out in the following manner. The bath is made cold with the required amount of dye-stuff and not too small a quantity of water, the cotton is immersed and worked for a short time to ensure impregnation, then the temperature is slowly raised to the boil. This operation should be carefully carried out, inasmuch as time is an important element in the dyeing with mordant colours; the colouring principle contained in the dye-stuff must enter into a chemical combination with the mordant that has been fixed on the fibre. Heat greatly assists this being brought about, but if the operation is carried on too quickly, then there is a tendency for uneven shades to be formed. This can only be remedied by keeping the temperature low until the dye-stuff has been fairly well united with the mordant, and then maintaining [Pg 162] the heat at the boil to ensure complete formation of the colouring lake on the fibre, and therefore the production of fast colours.

It has been noticed in the dyeing of alizarines on both cotton and wool that when, owing to a variety of circumstances, local overheating of the bath happens to take place dark strains or streaks are sure to be formed. To avoid these care should be taken that no such local heating can occur.

It only remains to add that it is possible to dye a great range of shades by this method, reds with alizarine and alumina; blacks with logwood and iron; greens from logwood, fustic, or Persian berries, with chrome and iron; blues from alizarine blues; greens from Coeruleine or Dinitrosoresorcine, etc.

Another method of mordanting cotton for the mordant group of dye-stuffs is that in which the cotton is impregnated with a salt of the mordant oxide derived from a volatile acid such as acetic acid, and then subjected to heat or steaming. This method is largely taken advantage of by calico printers for grounds, and dyers might make use of it to a much larger extent than they do.

There are used in this process the acetates of iron, chromium and aluminium, and bisulphites of the same metals and a few other compounds. Baths of these are prepared, and the cotton is impregnated by steeping in the usual way; then it is gently wrung out and aged, that is, hung up in a warm room overnight. During this time the mordant penetrates more thoroughly into the substance of the fabric, while the acid, being more or less volatile, passes off—probably not entirely, but at any rate some of the metal is left in the condition of oxide and the bulk of it as a basic salt. Instead of ageing the cotton may be subjected to a process of steaming with the same results. After this the [Pg 163] cotton is ready for dyeing, which is done by the method described in the last section.

There is still another method to be noticed here, that is, one in which a bath is prepared containing both the mordant and the dye-stuff. In this case the character of the mordant must be such that, under the conditions that prevail, it will not form a colour lake with the dye-stuff. Such substances are the bisulphites, if used with the bisulphite compounds of the dye-stuffs; the acetates, if mixed with some acetic acid, may also be used. The process consists in preparing the dye-bath containing both the mordant and the dye-stuff, entering the cotton, steeping for some time, then wringing and steaming. During the latter operation the acid combined with the mordant, being volatile, passes away, and the colouring matter and mordant enter into combination to form the colour lake, which is firmly fixed upon the fibre. Very good results may be obtained by this method.

Lastly, in connection with the mordant colours, attention may be directed to the process of using some of them, which consists in making a solution of the dye-stuff in ammonia, impregnating the cotton with this alkaline solution, and subjecting it to a steaming operation, during which the alkali, being volatile, passes away, leaving the colouring matter behind in an insoluble form. The cotton is next passed into a weak bath of the mordant (preferably the acetates of iron, etc.)., this being used first cold and then gradually heated up. The dye on the fibre and the mordant combine to form the desired colour, which is fixed on the fibre.

The chrome mordants are those which are most commonly applied by the methods here sketched out, and with the large and increasing number of mordant dyes available, the processes should be worth attention from the cotton dyer.

The following recipes give fuller details than the outline [Pg 164] sketches of the methods given above for the use of the various dyes produced with the mordant dyes and metallic mordants. In some cases as will be seen other dyes may be added to produce special shades: —

Dark Olive.—Prepare a bath from 8 lb. cutch, 4 lb. logwood extract, 7 lb. fustic extract, 2 lb. copper sulphate. Work in this for one to one and a half hours at the boil. This bath may be kept standing, adding new ingredients from time to time, and works best when it gets old. Then pass into a cold bath of 3 lb. copperas for one hour, then wash and enter into a new bath of 10 lb. salt, 6 oz. Titan blue 3 B, 6 oz. Titan brown R, 6 oz. Titan yellow Y, work for one hour at the boil, then lift, wash and dry.

Brown.—Prepare a bath with 20 lb. cutch, 2 lb. copper sulphate, 4 lb. quercitron extract. Work for one and a quarter hours at the boil, then allow to lie for a day, when the goods are passed into a bath containing 3 lb. bichromate of potash and 1 lb. alum. Work at 150° to 160° F. in this for a few minutes, then allow to lie for four to five hours, wash well and dry.

Olive.—Work for twenty minutes at 80° F. in a bath of 10 lb. fustic extract, 5 lb. quercitron extract, 2 lb. logwood extract; heat to boil, work for half an hour, then enter in a cold bath of 2 lb. sodium bichromate and 5 lb. copper sulphate; work for twenty minutes, then heat to boil; work for twenty minutes more, wash and dry.

Pale Brown. — Treat in a hot bath of 25 lb. cutch, 1¾ lb. bluestone; work for half an hour in this bath, then lift, wring, and work in a bath of 1¾ lb. bichromate of potash for twenty to thirty minutes. Dye in a bath of 2¼ lb. alum, 7 oz. Chrysoidin, 14 oz. Ponceau B.

Fast Brown. — The cotton is heated in a boiling bath containing 20 lb. cutch, 4 oz. copper sulphate for one hour, it is then treated in a bath containing 8 oz. bichromate of potash for half an hour, then dyed in a bath containing 2 oz. Benzo [Pg 165] black blue, 6 oz. Benzo brown N B, 2 lb. soap, 8 lb. salt, for one hour at the boil, washed and dried.

Drab. — Dissolve ½ lb. cutch, 7 lb. bluestone, 8 lb. extract of fustic; enter goods at 120° F., give six turns, lift and drain. Prepare a fresh bath containing 2 lb. copperas; enter goods, give three turns, lift, and enter fresh bath at 120°, containing 2 lb. bichromate of potash, give four turns, drain, wash and dry.

Coffee Brown. — For one piece, wet out in hot water, run for half an hour upon a jigger in a bath of 6 lb. good cutch, take up and drain in a bath of 8 lb. black iron liquor; drain, run again through each bath and rinse well. Prepare a fresh bath with Bismarck brown, enter at 100° F., heat slowly to 200° F., drain, rinse and dry.

Dark Brown Olive. — Prepare the dye-bath with 12 lb. cutch, 2 lb. bluestone, 2½ lb. alum, 10 lb. quercitron extract, 2 lb. indigo carmine 4 lb. turmeric, ¼ lb. Bismarck brown; boil for one and a half hours, then lift and add 1 lb. copperas; re-enter the goods, give another half-hour, boil, then add 1½ lb. bichromate of potash, work two hours more, then wash and dry.

Red Drab. — Boil up 10 lb. cutch and 5 lb. sumac; enter the cotton at 140° F., work fifteen minutes and lift. Prepare a fresh bath of 4 lb. black iron liquor; enter the cotton cold, work ten minutes and lift. Prepare another bath with 3 lb. bichromate of potash; enter cotton at 160° F., work fifteen minutes, lift and wash. Finish in a fresh bath containing 3 lb. logwood, 6 lb. red liquor; enter cotton at 100° F., work ten minutes, lift, wash and dry.

Fawn.—Boil up 5 lb. cutch and 5 oz. bluestone, cool to 100° F.; enter, give six turns, lift, and add 2 lb. copperas; re-enter cotton, give four turns, lift and wring. Prepare a fresh bath with 1 lb. bichromate of potash; enter cotton at 110° F., give five turns, lift, wash and dry. [Pg 166]

Grey Slate.—Boil up 10 lb. sumac, 3 lb. fustic extract; cool down to 120° F., give eight turns, lift and wring. Prepare a fresh bath with 5 lb. copperas; enter cotton cold, give five turns, lift and wash.

Dark Plum.—Lay down overnight in 30 lb. sumac. Next morning wring and enter in a fresh bath of oxy-muriate of tin 20° Tw., give four turns, lift and wash well in two waters. Boil out 40 lb. ground logwood, 10 lb. ground fustic, cool bath down to 140° F.; enter cotton, give eight turns, lift and add 1½ gallons red liquor; re-enter yarn, give four turns, lift, wash and dry.

Pale Chamois.—Work the cotton seven turns in a cold bath of 3 lb. copperas, then wring and pass into a cold bath of 3 lb. soda ash; work well, wash and dry.

Dark Brown Olive.—Prepare a bath of 28 lb. fustic, ¾ lb. logwood, 18 lb. cutch, 4 lb. turmeric, 2 lb. copper sulphate, ¾ lb. alum; work for an hour at the boil, then sadden in a new bath of 1 lb. bichromate of potash for half an hour, then sadden in a new bath of ¼ lb. nitrate of iron, working in the cold for half an hour, lift, wash and dry.

Havana Brown.—Prepare a bath with 4 lb. cutch and 1 lb. bluestone; work at the boil for one hour, then pass through a warm bath of ½ lb. bichromate of potash, 1 lb. sulphuric acid. Wash and dye in a bath of ¾ lb. Bismarck brown and 4 lb. alum; work for one hour at about 180° F., wash and dry.

Black.—Prepare a dye-bath with 20 lb. extract of logwood, 4 lb. cutch, 5 lb. soda ash, 5 lb. copper sulphate. Heat to the boil, enter the cotton, and work well for three hours, then lift, and allow to lie overnight in a wet condition, wash and pass into a bath of 1 lb. bichromate of potash for half an hour; lift, wash and dry. The dye-bath is not exhausted, and only about one-third of the various drugs need be added for further batches of cotton. [Pg 167]

Reseda Green.—Prepare a bath with 15 lb. cutch, 8 lb. turmeric; work in this for fifteen minutes at about 150° F., then pass through a hot bath of 2 lb. bichromate of potash for one hour, then re-enter into a cutch bath to which has been added, 1 lb. sulphate of iron; work for one hour, then add 2 lb. alum and work half an hour longer, rinse, wash and dry.

Fawn Brown.—Prepare a dye-bath with 4 lb. cutch, 2 lb. fustic extract; work for one hour at hand heat, then lift, and pass through a bath of 1¼ lb. bichromate of potash; work for a quarter of an hour, rinse and pass into a fresh bath of 1 oz. Bismarck brown for ten minutes, then lift, wash and dry.

Beige.—Prepare a bath with 20 lb. sumac; enter cotton at 120° F., give six turns, lift and add ½ lb. copperas; re-enter cotton, give four turns and wring. Prepare a fresh bath containing 2 lb. extract of fustic, 3 oz. extract of indigo; enter cotton at 120° F., give three turns, raise temperature to 140° F., and turn to shade, lift, wash and dry.

Turkey Red.—One of the most important colours dyed on cotton is that known as Turkey red, a bright red of a bluish tone, characterised by its great fastness to light, washing, etc. Strong alkalies turn it more yellowish, but weak acids and alkalies have little action.

Into the history of the dyeing of Turkey red it is not intended to enter, those who are interested in the subject should refer to old works on dyeing; nor is it intended to speak of old methods of producing it with the aid of madder, but rather to give some of the most modern methods for dyeing it with alizarine.

Many processes differing somewhat in detail have been devised for dyeing Turkey red on cotton, and it is probable that no two Turkey-red dyers work exactly alike. It is difficult to produce the most perfect red, and a very great [Pg 168] deal of care in carrying out the various operations is necessary to obtain it. This care and the number of operations makes Turkey red an expensive colour to dye, and so shorter methods are in use which dye a red on cotton that is cheaper, but not so brilliant or fast as a true Turkey red.

Process 1.—This process is perhaps the most elaborate of all processes, but it yields a fine red. The process is applicable to cloth or yarn, although naturally the machinery used will vary to suit the different conditions of the material. Bleached yarn or cloth may be treated, although a full bleach is not necessary, but the cloth or yarn must be clean or well scoured, so that it is free from grease and other impurities.

Operation 1. Boil the cotton for six to eight hours with a carbonate of soda lye at 1° Tw. in a kier at ordinary pressure, then wash well, wring, or, better, hydro-extract.

Operation 2. First "greening": What is called the "first green liquor" is prepared by taking 15 lb. of gallipoli oil, 3 lb. phosphate of soda and 15 lb. carbonate of soda, the liquor to stand at 2° Tw. Originally this "liquor" was made with sheep dung, but this is now omitted. The cotton is worked in this liquor, which is kept at 100° F., until it is thoroughly impregnated, then it is taken out, squeezed and dried, or in some cases piled overnight and then stoved.

Operation 3. Second green liquor. As before.

Operation 4. Third green liquor. As before.

Operation 5. A carbonate of soda liquor of 2° Tw. strength is prepared, and the cotton steeped in this until it is thoroughly impregnated, then it is wrung out and stoved. This is called "white liquor treatment".

Operation 6. Second white liquor. As before.

Operation 7. Steeping: Prepare a bath of water at 150° F., and steep for twelve hours, then wring and dry.

Operation 8. Sumacing: A liquor is made from 12 lb. [Pg 169] sumac with water, and after straining from undissolved sumac leaves the liquor is made to stand at 2° Tw., this is kept at about a 100° F., and the cotton is well worked in it and allowed to steep for four hours, after which it is taken out and wrung.

Operation 9. Mordanting or aluming: 20 lb. of alum are dissolved in hot water, and 5 lb. of soda crystals are slowly added in order to prepare a basic alum solution; this is now made by the addition of water to stand at 8° Tw.

The sumaced cotton is worked in this bath and allowed to steep for twenty-four hours, when it is taken out and wrung. Some dyers add a little tin crystals to this bath; others add a small quantity of red liquor.

Operation 10. The dyeing: A cold bath is prepared with 10 lb. to 12 lb. alizarine, 3 lb. sumac extract, and 2 oz. lime. The cotton is entered into the cold bath, worked from fifteen to twenty minutes so as to get it thoroughly impregnated; then the heat is slowly raised to the boil and the dyeing carried on at that heat until the full shade is obtained, which usually takes about an hour. According to the brand of alizarine used so will the shade that is obtained vary, as will be mentioned later on.

Operation 11. First clearing: The dyed cotton is placed in a boiler and boiled for four hours with 3 lb. soda crystals and 3 lb. palm oil soap, afterwards washing well.

Operation 12. Second clearing: The dyed cotton is again boiled for two hours with 2½ lb. soap and ½ lb. tin crystals, then give a good washing and dry.

This process is a long one—indeed, some dyers by repeating some of the operations lengthen it—and it takes at least two weeks, in some cases three weeks, to carry out.

The first idea is to get the cotton thoroughly impregnated with the oil, and this oxidised to some extent on the fibre, and to this end the oil treatments are carried out. In this [Pg 170] process experience has shown that olive oil is the best to use, although other oils have been tried from time to time. The sumacing enables the alumina to be more firmly fixed on to the cotton. The alumina combines with both the oil and the sumac, and the resulting mordant produces a better and more brilliant red with the alizarine. The clearing operations serve to remove impurities, to brighten the colour, and to more fully fix it on the cotton.

Process 2.—Operation 1. The cotton is well bleached or scoured with soda in the usual way.

Operation 2. Oiling or preparing: A liquor is made from 10 lb. alizarine oil or Turkey-red oil in 10 gallons water. This oil is prepared from castor oil by a process of treatment with sulphuric acid, washing with water and neutralising with caustic soda. The cotton is thoroughly impregnated with this oil by steeping, then it is wrung out and dried.

Operation 3. Steaming: The cotton is put into a steaming cottage or continuous steaming chamber and steamed for from one to one and a half hours at about 5 lb. pressure.

Operation 4. A bath of red liquor (acetate of alumina) at 8° Tw. is prepared. Some dyers use basic alum at the same strength. In this bath the cotton is steeped at 100° F. for two hours; then it is wrung out and dried. This aluming bath can be repeated. Next it is run through a bath of chalk and water containing 2 lb. chalk in 10 gallons water. This helps to fix the alumina on the cotton. Phosphate of soda also makes a good fixing agent.

Operation 5. Dyeing: This is carried out in precisely the same way as in the other process.

Operation 6. Oiling: A second oiling is now given in a bath of 5 lb. alizarine oil, or Turkey-red oil, in 10 gallons water, after which the cotton is dried, when it is ready for [Pg 171] further treatment. In place of giving a second oiling after the dyeing, it is, perhaps, better to give it after the mordanting and before dyeing.

Operation 7. Clearing: The dyed cotton is cleared with soap in the same manner as the clearing operations of the first process, which see.

Any of the treatments preparatory to, and following the actual dyeing of, any of these processes may be repeated if deemed necessary. The text-books on dyeing and the technical journals devoted to the subject frequently contain accounts of methods of dyeing Turkey red, but when these come to be dissected the methods are but little more than variants of those which have just been given.

Seeing that the theory or theories involved in this rather complex process of dyeing Turkey red, and that colourists are not agreed as to the real part played by the oil, the sumac and the clearing operations in the formation of a Turkey red on cotton, nothing will be said here as to the theory of Turkey-red dyeing.

Alizarine Red.—It is possible to dye a red with alizarine on cotton which, while being a good colour, is not quite so fast to washing, etc., as a Turkey red. This is done by using fewer treatments, as shown in the following process:—

Process 1.—Boil the cotton in soda.

Process 2.—Oil with Turkey-red oil, as in the Turkey-red process No. 2 above.

Process 3.—Mordant with alum or acetate of alumina.

Process 4.—Dye with alizarine as before.

Process 5.—Soap.

There are three distinct colouring matters which are sold commercially under the name of "alizarine". These are: alizarine itself, which produces a bluish shade of red; anthra-purpurine, which gives a similar but less blue red than alizarine; and flavo-purpurine, which produces the yellowest reds. The makers send out all these various products under various marks. [Pg 172]

For dyeing Turkey reds the flavo-and anthra-purpurine brands or yellow alizarines are to be preferred; for pinks and rose shades the alizarine or blue shade brands are best.

Alizarine Pink.—This can be dyed in the same way as Turkey red, only using for full pinks 4 per cent, of alizarine in the dye-bath, or for pale pinks 1 to 2 per cent. It is advisable to reduce the strength of the oiling and mordanting baths down to one-half.

Alizarine Violet.—Alizarine has the property of combining with iron to form a dark violet colour, and advantage is taken of this fact to dye what are called in the dyeing and calico printing trades alizarine purples and lilacs, although these do not resemble in hue or brilliance the purples and lilacs which can be got from the direct dyes. They have not the importance which they formerly possessed, and but a mere outline of two processes for their production will be given.

Alizarine Purple.—*Process* 1. (1) Boil with soda, (2) prepare with Turkey red oil, (3) mordant by steeping in copperas liquor at 4° Tw. for twenty minutes, take out, allow to lie on stillages overnight, then wash and dry. For deep purples it may be advisable to repeat these treatments; for pale lilacs using them at half strength is advisable. (4) Dye with 8 to 10 per cent. of alizarine blue shade, working as described under Turkey red. The best results are obtained when 1 per cent, of chalk is added to the dye-bath. (5) Soap as in red dyeing.

Process 2. (1) Boil with soda, (2) oil with Turkey-red oil, (3) steep in pyrolignite of iron (iron liquor) for one hour, then age by hanging in the air. (4) Dye as before. (5) Soap.

Fine blacks are got if after oiling the cotton is treated with sumac or tannic acid, then mordanted with iron and dyed with alizarine as usual.

Chocolate Browns.—Fine fast chocolate browns can be [Pg 173] got from alizarine by using a mixed mordant of iron and alumina, either the acetate or the sulphate. By varying the relative proportions various shades can be obtained.

Alizarine Orange—Prepare the cotton as if for dyeing a Turkey red, but use in the dye-bath 8 to 10 per cent. of Alizarine orange.

Alizarine Blue—The cotton is boiled three hours with 3 per cent. ammonia soda at 30 lb. pressure, and then washed thoroughly. The boiled, washed and hydro-extracted yarn is oiled with a solution containing from ¼ lb. to 1½ lb. Turkey-red oil, 50 per cent. for every gallon of water. It is then wrung out evenly and dried for twelve hours at 150° F.

Tannin Grounding.

The oiled and dried cotton is worked three-quarters of an hour in a vat containing a tannin solution (1 oz. per gallon). The cotton remains in this liquid, which is allowed to cool off for twelve hours, then it is hydro-extracted. Sumac turns the shade somewhat greener, which is noticed especially after bleaching, therefore tannin is given the preference.

Chromium Mordant.

The cotton treated with tannin and then hydro-extracted is worked cold for one hour in a vat containing a solution of chromium chloride at 32° Tw., and remains in this solution twelve hours. The cotton is then hydro-extracted and washed directly; it is best to employ running water. A special fixation does not take place. The cotton is now ready for dyeing. The solution of chromium chloride and the tannin solution can be used continuously, adding fresh liquor to keep the baths up to strength.

Dyeing—For dyeing, water free from lime must be used. Water having not more than 2.5° hardness can be employed [Pg 174] if it is corrected with acetic acid, thereby converting the carbonate of lime into acetate of lime. Very calcareous water must be freed from lime before use. The dye-bath contains for 100 lb. cotton 15 lb. Alizarine blue paste (A R or F, according to the shade desired), 35 lb. acetic acid (12° Tw.), 15½ lb. ammonia (25 per cent.), 2¼ oz. tannin. The cotton is worked a quarter of an hour in the cold; the temperature is raised slowly to a boil, taking about one hour, and the cotton is worked three-quarters of an hour at that heat. Finally the cotton is washed and hydro-extracted. The dyed and washed cotton is steamed two hours at 15 lb. to 22 lb. pressure. Steaming turns the shade greener and darker, and increases the fastness. After steaming the cotton it is soaped one or two hours at the boil, with or without pressure. According to the quality of water employed, 2 to 5 parts soap per 1,000 parts water are taken.

Brown.—A fine brown is got by a similar process to this, if instead of Alizarine blue, Alizarine orange is used in the dye-bath. A deeper brown still if Anthracene brown, or a mixture of Anthracene brown and Alizarine blue, be used.

Claret Red.—Clarets to maroon shade of red are got by preparing the cotton as for blue given above, then dyeing with alizarine.

Logwood Black.—One of the most important colours that come under this section is logwood black, the formation of which on the fibre depends upon the fact that the colouring principle of logwood forms a black colour lake with iron and also one with chromium.

There are many ways of dyeing logwood blacks on cotton, whether that be in form of hanks of yarns, warps or pieces. While these blacks may be, and in the case of hanks are, dyed by what may be termed an intermittent process, yet for warps and piece goods a continuous process is preferred by dyers. Examples of both methods will be given. As in the dyeing of Turkey reds it is probable that no two dyers [Pg 175] of logwood blacks quite agree in the details of their process, there may be variations in the order of the various baths and in their relative strengths. Typical methods will be noted here.

Dyeing Logwood Black on Yarn in Hanks.—Operation 1. Sumacing: Prepare a bath with 10 lb. sumac extract in hot water. Work the yarn in this for half an hour, then allow to steep for six hours or overnight, lift and wring. The liquor which is left may be used again for another lot of yarn by adding 5 lb. sumac extract for each successive lot of yarn. In place of using sumac the cheaper myrabolam extract may be used.

Operation 2. Ironing or Saddening: Prepare a bath with 3½ gallons nitrate of iron, 80° Tw. Work the yarn in this for fifteen minutes, then wring out. The bath may be used again when 1 gallon of nitrate of iron is added for each lot of yarn worked in it. In place of the nitrate of iron, the pyrolignite of iron or iron liquor may be used.

Operation 3. Liming: Work for ten minutes in a weak bath of milk of lime.

Operation 4. Dyeing: This is done in a bath made from 10 lb. logwood extract and 1 lb. fustic extract. The yarn is entered into the cold or tepid bath, the heat slowly raised to about 150° F, then kept at this heat until a good black is got, when the yarn is taken out, rinsed and wrung. The addition of the fustic extract enables a much deeper and jetter shade of black to be dyed.

Operation 5. Saddening: To obtain a fuller black the dyed cotton is sent through a bath of 1½ lb. of copperas, then washed well.

Operation 6. Soaping: Work for twenty minutes in a bath of 2 lb. soap at 140° to 150° F. Then wash well.

Much the same process may be followed for dyeing [Pg 176] logwood black on warps and piece goods, jiggers being used for each operation.

Another method is to first work the cotton in pyrolignite (iron liquor) at 10° Tw., until it is thoroughly impregnated, then to dry and hang in the air for some hours, next to pass through lime water to fix the iron, and then to dye as before.

Continuous Process. — In this case a continuous dyeing machine is provided, fitted with five to six compartments. The cotton is first of all prepared by steeping in a bath of 12 lb. myrabolam extract for several hours, then it is taken to the continuous machine and run in succession through nitrate of iron liquor, lime water, logwood and fustic, iron liquor and water. The nitrate of iron bath contains 2 gallons of the nitrate to 10 gallons of water, and as the pieces go through fresh additions of this liquor are made from time to time to keep up the volume and strength of the liquor to the original points.

The logwood bath is made from 10 lb. logwood extract and 1 lb. fustic extract, and it is used at about 160° F. The quantities here given will serve for 100 lb. of cotton, and it is well to add them dissolved up in hot water in small quantities from time to time as the cotton goes through the bath.

The iron liquor given after the dyeing contains 2 lb. of copperas in 10 gallons of water.

Between the various compartments of the machine is fitted squeezing rollers to press out any surplus liquor, which is run back into the compartment. The rate of running the warp or pieces through should not be too rapid, and the dyer must adapt the rate to the speed with which the cloth dyes up in the dye-bath.

The addition of a little red liquor (alumina acetate) to the iron bath is sometimes made, this is advantageous, as it results in the production of a finer black. Iron by itself [Pg 177] tends to give a rusty-looking, or brownish black, but the violet, or lilac shade that alumina gives with logwood, tones the black and makes it look more pleasant.

Some dyers add a small quantity, 1 per cent., of the weight of the cotton of sulphate of copper to the iron bath, others add even more than this. Some use nitrate of copper; the copper giving a greenish shade of black with logwood, and this tones down the iron black and makes it more bloomy in appearance.

Single bath methods of dyeing logwood blacks are in use, such methods are not economical as a large quantity, both of dye-wood and mordants, remain in the bath unused. Although full intense blacks can be dyed with them, the black is rather loosely fixed and tends to rub off. This is because as both the dye-stuff and the mordant are in the same bath together they tend to enter into combination and form a colour lake that precipitates out in the dye-bath, causing the loss of material alluded to above, while some of it gets mechanically fixed on the cotton, in a more or less loose form, and this looseness causes the colour to rub off.

For a *chrome-logwood black*, a dye-bath is made with 3 lb. bichromate of potash, 100 gallons logwood decoction at 3° Tw., and 6½ lb. hydrochloric acid. Enter the cotton into the cold bath, raise slowly to the boil and work until the cotton has acquired a full black blue colour, then take it out and rinse in a hot lime water when a blue black will be got.

A *copper-logwood black* is got by taking 100 gallons logwood decoction at 3° Tw., and 6 lb. copper acetate (verdigris); the cotton is entered cold and brought up to the boil. Copper nitrate may be used in the place of the copper acetate, when it is a good plan to add a little soda to the bath. Some dyers in working a copper-logwood black make the dye-bath from 100 gallons logwood liquor at 2° Tw., 4 lb. [Pg 178] copper sulphate (bluestone) and 4 lb. soda. This bath is used at about 180° to 190° F., for three-quarters of an hour, then the cotton is lifted out, wrung and aged or as it is sometimes called "smothered" for five hours. The operations are repeated two or three times to develop a full black.

Logwood black dyeing has lost much of its importance of late years owing to the introduction of the many direct blacks, which are much easier of application and leave the cotton with a fuller and softer feel.

Logwood Greys.—These are much dyed on cotton and are nothing more than weak logwood blacks, and may be dyed by the same processes only using baths of about one-tenth the strength.

By a one-bath process 5 lb. of logwood are made into a decoction and to this 1 lb. of copperas (ferrous sulphate) is added and the cotton is dyed at about 150° F. in this bath. By adding to the dye-bath small quantities of other dye-woods, fustic, peach wood, sumach, etc., greys of various shades are obtained. Some recipes bearing on this point are given in this section.

Logwood is not only used for dyeing blacks and greys as the principal colouring matter, but is also used as a shading colour along with cutch, fustic, quercitron, etc., in dyeing olives, browns, etc., and among the recipes given in this section examples of its use in this direction will be found.

The dye-woods—fustic, Brazil wood, bar wood, Lima wood, cam wood, cutch, peach wood, quercitron bark, Persian berries—have since the introduction of the direct dyes lost much of their importance and are now little used. Cutch is used in the dyeing of browns and several recipes have already been given. Their production consists essentially in treating the cotton in a bath of cutch, either alone or for the purpose of shading with other dye-woods when the cotton takes up [Pg 179] the tannin and colouring matter of the cutch, etc. The colour is then developed by treatment with bichromate of potash, either with or without the addition of an iron salt to darken the shade of brown.

The usual methods of applying all the other dye-woods, to obtain scarlets to reds with Brazil wood, Lima wood, peach wood; or yellows with fustic, quercitron or Persian berries, is to first prepare the cotton with sumac, then mordant with alumina acetate or tin crystals (the latter gives the brightest shades), then dye in a decoction of the dye-woods. Sometimes the cotton is boiled in a bath of the wood when it takes up some of the dye-wood, next there is added alumina acetate or tin crystals and the dyeing is continued when the colour becomes developed and fixed upon the cotton.

Iron may be used as a mordant for any of these dye-woods but it gives dull sad shades.

Chrome mordants can also be used and these produce darker shades than tin or alumina mordants.

As practically all these dye-woods are now not used by themselves it has not been deemed necessary to give specific recipes for their application, on previous pages several are given showing their use in combination with other dyes.

The dye-stuff Dinitroso-resorcine or Solid green O is used along with iron mordants for producing fast greens and with chrome mordants for producing browns to a limited extent in cotton dyeing. The following recipes give the details of the process.

Green.—Steep the cotton yarn or cloth in the following liquor until well impregnated, then dry: 3 gallons iron liquor (pyrolignite of iron), 22° Tw. gallons of water, ¾ gallon acetic acid, 12° Tw., 2 lb. ammonium chloride. Then pass the cotton through a warm bath of 3 oz. phosphate of soda and 4 oz. chalk per gallon, then enter into a dye-bath containing 6 lb. Solid green O. Work as described for dyeing [Pg 180] alizarine red. For darker greens of a Russian green shade use 10 lb. of solid green O, in the dye-bath.

Brown.—A fine brown is got by steeping the cotton in a bath of 8 lb. Solid green O, 6¾ gallons water, 1½ gallons ammonia and 2 lb. acetate of chrome; dry, then pass through a soap-bath, wash and dry.

Deep Olive Brown.—Mix 8 lb. Solid green O and 4½ lb. borax with 6 gallons water, add ½ lb. Turkey-red oil, 5 lb. ammonia, then 2 gallons water and 1½ lb. copper-soda solution and another 2 gallons water. Steep the cotton in this, dry, soap well and wash. The copper-soda solution is made from 10 lb. chloride of copper (75° Tw.), 5 lb. tartaric acid, 12 lb. caustic soda (75° Tw.) and 4 lb. glycerine.

Khaki.—Make the dye liquor from 14 lb. Solid green O, ½ lb. Alizarine yellow N, 1 lb. caustic soda (36° Tw.), ½ lb. Turkey-red oil and 8 gallons water. To this add 2½ lb. acetate of chrome (32° Tw.), 2¼ lb. copper-soda solution and 4 gallons water.

Sage Green.—Use 1¼ lb. Solid green O, 3 lb. caustic soda (36° Tw.), ½ lb. Ceruleine, ½ lb. Turkey-red oil, 1 gallon water to which is added 2½ lb. acetate of chrome (32° Tw.) and 2¼ lb. copper-soda solution dissolved in 4 gallons water.

Pale Brown.—Use 4 lb. Solid green O, 2½ lb. borax, 3 lb. ammonia, ½ lb. Turkey-red oil, 6 gallons of water and 1½ lb. copper-soda solution dissolved in 2 gallons water.

Pale Fawn Brown.—The dye-bath is made from ½ lb. Alizarine, 1¼ lb. Solid green O, 1½ lb. borax, ½ lb. Turkey-red oil and 5 gallons of water to which is added 1½ lb. acetate of chrome (32° Tw.), 1½ lb. copper-soda solution and 4 gallons water. In all cases the cotton is steeped in the dye liquors until thoroughly impregnated, then the excess liquor is wrung out, the cotton dried, then passed through a soap bath, washed well and dried.

Dark Brown. — Place the cotton in a lukewarm bath of 25 [Pg 181] lb. cutch and 1½ lb. copper sulphate; work for half an hour, then steep for six hours, then lift, wring and enter into a bath of 3¼ lb. bichromate of potash at 160° F. for twenty minutes. Then wash and dry.

Yellow Brown. — Make a bath with 14 lb. cutch and ½ lb. copper sulphate; work in this bath for four hours at 120° F., then pass into a bath of 2 lb. copperas and ½ lb. chalk, work for half an hour in the cold, then pass into a hot bath of 2½ lb. bichromate of potash at 150° F. for half an hour.

Dark Brown. — Make a dye-bath with 15 lb. cutch, 2 lb. logwood extract and 2 lb. fustic extract; work the cotton in this at 160° F. for three hours, then pass into a cold bath of 1 lb. copperas and ¼ lb. chalk for half an hour, then into a bath of 3 lb. bichromate of potash for half an hour at 150° F., then wash and dry.

(7) PRODUCTION OF COLOUR DIRECT UPON COTTON FIBRES.

By the action of nitrous acid upon the salts of the primary organic amines the so-called diazo compounds are formed. An example of this important process is that of nitrous acid on aniline hydrochloride shown in the following equation: —

$C_6H_5NH_2$ + HCl + HNO_2 $2H_2O$ + $C_6H_5N{:}NCl$ Hydrochloric acid Nitrous Water, Diazo-benzene aniline, acid, chloride.

These diazo compounds are distinguished by their active properties, especially in combining with amines in acid solutions, or with phenols in alkaline solution to form the azo dyes, thus diazobenzene chloride will combine with naphthol to form naphthol-azo-benzene, thus: —

$C_6H_5N{:}NCl$ + $C_{10}H_7OH$ + NaOH = Diazo-benzene chloride, Naphthol, Caustic soda.

$C_{10}H_6OHN{:}NC_6H_5$ + NaCl + H_2O Naphthol-azo-benzene, Salt, Water. [Pg 182]

These azo compounds are coloured, but are perfectly insoluble in water, alkalies, or acids; on the other hand the sulphonates of these bodies are easily soluble and form the numerous azo dyes now so largely made and used in wool and silk dyeing, but which on account of their being sulphonates cannot be used for cotton dyeing.

Methods have been devised for producing the insoluble azo colours direct upon the fibres. They are also called naphthol colours from the use of beta-and alpha-naphthol in their production. Although these azo dyes, when produced on the fibre, do not possess the fastness of the alizarine dyes, yet, on account of their cheapness and relative great fastness to soap and the action of sunlight, they are better than many of the newer cotton dyes.

By this method (first introduced in England by Holliday) colours of exceptional brightness and fastness can be obtained which were not obtainable with the dyes then known. Those which are obtained from phenols are of the first importance.

The Diazotisation of the Amido Bases.

With most bases this must be accomplished as cold as possible below 65° F. At a higher temperature, and when allowed to stand, most diazo compounds decompose quickly with evolution of nitrogen, which decomposition results in the mixture losing its power of producing colour, or at the most gives unsatisfactory results. For this reason it is therefore always necessary to work as cold and as quickly as possible.

The amido-azo bodies, whose compounds with the phenols are also distinguished by their great fastness, are in this respect an exception. They can be diazotised at the ordinary temperature, and their diazo compounds are much stabler than those, for example, of alpha-and beta-naphthylamine or of aniline, which must always be used as quickly as possible. [Pg 183]

From anisidine, phenetidine and amido-diphenylamine, still more stable diazo compounds can be obtained, but the prices of these bases are rather high, and the colours produced with them are not fast to light.

The cheapest and most convenient method of obtaining nitrous acid for diazotising is by the action of a mineral acid, preferably hydrochloric acid, upon nitrite of soda.

For diazotising one molecule of base requires one molecule of hydrochloric acid to form a salt of the base, a molecule of nitrite of soda, and another molecule of hydrochloric acid to decompose the nitrite. The diazotisation is better carried out and the diazo solution rendered more stable if another molecule of hydrochloric acid and an excess of nitrite of soda are used. The presence of an excess of nitrite can be determined by testing the diazo solution with potassium iodide starch paper, which in the presence of excess of nitrite gives the blue iodine starch reaction.

In carrying out the diazotisation, the base is first dissolved in the whole amount of hydrochloric acid which has to be used, and the solution is filtered. The diazotisation takes place in the manner shown in the equation: —

$C_6H_5NH_2$ + HCl +   HCl   + $NaNO_2$ =
Aniline hydrochloride, Hydrochloric acid, Sodium nitrite,
NaCl   + $C_6H_5N{:}NCl$ + H_2O
Salt,   Diazo-benzene chloride,   Water.

The bases which form salts soluble with difficulty, such as nitroaniline and the amido-azo bodies, offer special difficulties in diazotising.

It has been found that the operation with these is best carried out if the chemically pure bases in paste form are mixed with the requisite amount of nitrite, and the diluted paste then poured into the hydrochloric acid. [Pg 184]

It has been found by experience that the colour is developed much brighter upon the fibre when the diazo solution contains acetic acid and no free mineral acid. However, the diazotisation is better carried out with hydrochloric acid, and the presence of the latter is necessary to give stability to the solution. If before the diazo solution is used a quantity of acetate of soda be added to it, the free hydrochloric acid liberates acetic acid from the acetate, and the chloride of the diazo body changes into its acetate. It is better to add an excess above the two molecules of acetate of soda which are required.

The combination when aniline and beta-naphthol are used, as the amine and phenol respectively, is shown in the following equations:—

$C_6H_5N{:}NCl$ + $C_{10}H_7OH$ +
Diazo-benzene chloride, B. naphthol,

$NaOH = 2NaCl + C_6H_5N{:}NC_{10}H_6OH + H_2O$ Caustic soda, Benzene- azo-naphthol, Water.

Or, with naphthylamine and naphthol, thus:—

$C_{10}H_7N{:}NCl + C_{10}H_7OH + NaOH =$

$NaCl$ + $C_{10}H_7N{:}NC_{10}H_6OH$ + H_2O
Naphthalene azo-naphthol.

By the action of nitrous acid upon amido-azo bodies a group of bodies called diazo-azo compounds are obtained which contain the group N:N twice over, thus:—

$C_6H_5N{:}NC_6H_4NH_2HCl + NaNO_2 + 2HCl =$
Benzene-azo-aniline-hydrochloride,
$NaCl + C_6H_5N{:}NC_6H_4N{:}NCl + 2H_2O.$

[Pg 185]

Diazo-azo-benzene-chloride.

When this compound is combined with naphthol diazo-azo dyes are produced.

$C_6H_5N{:}NC_6H_4N{:}NC_{10}H_6OH.$ Benzene-azo-benzene-azo-naphthol.

The molecular weights of the bases, phenols and chemicals employed are the following:—

1. Hydrochloric acid, $HCl-36.5$. 2. Caustic soda, $NaOH-40$. 3. Nitrite of soda, $NaNO_2-69$. 4. Acetate of soda, $NaC_2H_3O_23H_2O-136$.

176

1. Commercial hydrochloric acid at 32° Tw. contains about 365 grams of HCl in a litre, or 3½ lb. in a gallon.

2. The commercial 77 per cent. soda must always be used, and for practical purposes it may be taken as pure. It is best to make a solution which contains 160 grams NaOH in a litre of water.

3. The nitrite supplied is almost chemically pure, and is easily soluble in water. In order to make a solution 140 or 290 grams are dissolved per litre.

4. Crystallised acetate of soda contains 3 molecules of water of crystallisation, and is usually somewhat moist. Instead of 136 grams 140 are taken to allow for moisture. The amount is dissolved in about 500 cubic centimetres of water.

Bases.

1. Aniline, $C_6H_5NH_2 - 93$.
2. Toluidine, $C_7H_7NH_2 - 107$.
3. Alpha-and beta-naphthylamine, $C_{10}H_7NH_2 - 143$.
4. Para-or meta-nitroaniline, $C_6H_4NO_2NH_2 - 138$.
5. Nitro-para-toluidine, $C_7H_6NO_2NH_2 - 152$.
6. Amidoazobenzene (base), [Pg 186] $C_6H_5N{:}NC_6H_4NH_2 - 197$.
7. Orthoamidoazotoluol (base), $C_7H_7N{:}NC_7H_6NH_2 - 225$.
8. Alpha-or beta-naphthol $C_{10}H_7OH - 144$.

Example of Quantities Taken.

Molecular Weight.
1. Molecule nitrite 69 grams.
2. Molecule aniline 93 "
3. Molecule hydrochloric acid 365 "
4. Molecule acetate of soda 136 "
5. Molecule of naphthol 144 "
6. Molecule caustic soda 40 "

Applying the principles which have just been described to the dyeing of cotton, it is found that the cotton may be dyed by taking the base and preparing the diazo body, impregnating the cotton with this, and developing the colour by passing into a bath of the phenol. On the other hand, the cotton can be prepared with the phenol and the colour developed by passing into a bath of the diazotised base, and practice has shown that this latter proceeding is the best. Practically the only phenol that is used is the beta-naphthol; alpha-naphthol is occasionally used, but not often.

The purer the beta-naphthol the better, especially for producing the paranitroaniline red. Various preparations of beta-naphthol have been brought out by colour makers.

The process of dyeing cotton with a naphthol colour takes place in two stages, the first being the grounding or preparing with the naphthol, the second the developing with the diazotised base. Some of the effects which can be obtained from the two naphthols and various bases are given in the following table: —

Base.   With beta-naphthol, gives With alpha-naphthol, gives
1. Aniline, Orange yellow; Cutch brown.
2. Paratoluidine, Full yellow orange; Cutch brown.
3. Metanitroaniline, Fiery yellowish red; Brownish orange.
4. Paranitroaniline, Bright scarlet; " "

[Pg 187] Base.   With beta-naphthol, gives With alpha-naphthol, gives
5. Nitroparatoluidine, Orange; Very bright catechu.
6. Alpha-naphthylamine, Bluish claret red; Reddish puce.
7. Beta-naphthylamine, Turkey red; "
8. Amidoazobenzene, Red; "
9. Orthoamidoazotoluene, Yellowish claret red; "

By mixing alpha-and beta-naphthols together a variety of grenat and claret reds and browns can be obtained.

With regard to the fastness of the shades produced the following may be considered: —

Fast to Soaping.
Combination of A-Naphthol with Toluidine.
"   "   A-Naphthylamine.
"   "   B-Naphthylamine.
"   "   Amidoazobenzene.
"   B-Naphthol with Toluidine.
"   "   Paranitroaniline.

178

"   "   Nitroparatoluidine.
"   "   B-Naphthylamine.
"   "   A-Naphthylamine.

Moderately Fast.

Combination of A-Naphthol with Aniline.
"   "   Paranitroaniline.
"   "   Orthoamidoazotoluene.
"   B-Naphthol with Metanitroaniline.
"   "   Amidoazobenzene.

Very Loose.

Combination of A-Naphthol with Paratoluidine.
"   "   Metanitroaniline.
"   "   Nitroparatoluidine.
"   B-Naphthol with Aniline.
"   "   Paratoluidine.

[Pg 188]

"   "   Orthoamidoazotoluene.

The samples were tested for fastness to light by exposing them for nine days with the following results: —

Fast.

Combination of A-Naphthol with Aniline.
"   "   Toluidine.
"   "   Metanitroaniline.
"   "   Paranitroaniline.
"   "   Nitroparatoluidine.
"   "   B-Naphthylamine.
"   "   Amidoazobenzol.
"   "   Orthoamidoazotoluol.
"   B-Naphthol with Aniline.
"   "   Paratoluidine.
"   "   Metanitroaniline.
"   "   Paranitroaniline.
"   "   B-Naphthylamine.
"   "   A-Naphthylamine.

Moderately Fast.

Combination of B- Naphthol with Nitroparatoluidine.

Very Loose.
Combination of A-Naphthol with Toluidine.
"   "   A-Naphthylamine
"   B-Naphthol with Toluidine.
"   "   Amidoazobenzene.
"   "   Orthoamidoazotoluene.

The most important of the naphthol colours is undoubtedly para-nitroaniline red, produced by the combination of paranitroaniline and beta-naphthol. In order to produce the best and brightest shades these two bodies must be quite pure. [Pg 189] The following directions may be followed:—

Dyeing Paranitroaniline Red on Yarn.

It unfortunately happens that this red does not admit of being worked in large quantities at a time, particularly in the diazo bath where the colour is developed, as the previous operations seem to render the yarn slightly waterproof, and hence if large quantities of yarn were dealt with at one time some would be found to be dyed all right, others would be defective. It has, therefore, been found best to work only about 2 lb. of yarn at a time, carefully carrying out each operation with this quantity. As, however, the process can be quickly worked it follows that in the course of a day a fairly large quantity of yarn can be treated.

1. *Grounding.* The grounding or preparing bath for 100 lb. of yarn is best made in the following manner: 4 lb. of beta-naphthol are stirred in 2½ lb. of caustic soda liquor 70° Tw., then 1½ quarts of boiling water is added, when dissolved 1½ quarts of cold water. In a separate vessel dissolve 5 lb. Turkey-red oil in 11 quarts of water, then mix the two liquors together and add sufficient water to make up the whole to 12 gallons.

In working sufficient of this liquor is taken and put into a deep tub in which 2 lb. of yarn can be conveniently worked. It is best to work at a tepid heat, say 100° to 110° F.; 2 lb. of the yarn are worked in this liquor, so that it becomes thoroughly impregnated, then it is gently wrung out and hung up. This operation is repeated with each 2 lb. until the whole 100 lb. has been treated, adding from time to time some of the naphthol liquor to make up for that taken up by the cotton. When all the yarn has been through the liquor, give it another dip through the same liquor. Place the yarn in a hydro-extractor for five to seven minutes. Next open out the yarn well, and hang on sticks and dry in a stove at 140° to 150° F. The stove should be [Pg 190] heated with iron pipes, through which steam at 30 lb. to 40 lb. pressure passes. This stove should be reserved entirely for this work, for if other goods be dried in it along with the naphthol-prepared cotton, any steam or acid vapours which might be given off from the former might damage the latter.

When thoroughly dry the yarn is ready for the next operation.

2. *Developing*. The developing bath is made in the following manner: 1½ lb. paranitroaniline is mixed with 1½ gallons of boiling water, and 1¾ quarts of hydrochloric acid at 30° to 32° Tw. Stir well until the paranitroaniline is completely dissolved, add 3½ gallons of cold water, which will cause a precipitation of the hydrochlorate of paranitroaniline as a yellow powder. Let the mixture thoroughly cool off, best by allowing to stand all night; 1¼ lb. of nitrite of soda is dissolved in 4 quarts of cold water, and this solution is added to the paranitroaniline solution slowly and with constant stirring; in about fifteen to twenty minutes the diazotisation will be complete. At this and following stages the temperature of working should be kept as low as possible. Some dyers use ice in preparing their diazo solutions, and certainly the best results are attained thereby, but with paranitroaniline the ice can be dispensed with. After the end of the time sufficient cold water is added to bring the volume of the liquor up to 10 gallons. This diazo liquor will keep for some days, but it decomposes in time, so that it should not be kept too long.

Another liquor is made by dissolving 4 lb. acetate of soda in 11 quarts of water.

The developing bath is made by taking 4 gallons of the diazo liquor and 1 gallon of the acetate liquor and mixing together, and in this bath the prepared yarn, 2 lb. at a time, [Pg 191] is worked. The colour develops immediately. The yarn when dyed is lifted out, wrung, and then it is well washed with water, soaped in a bath at 120° F., with a liquor containing ½ oz. soap per gallon, then dried. As the cotton yarn is being passed through the developing bath, the latter is freshened up from time to time by suitable additions of the diazo and acetate liquors in the proportions given above.

Some dyers use a special form of dye vat for dyeing paranitroaniline red on yarn, whose construction can be seen from Fig. 27.

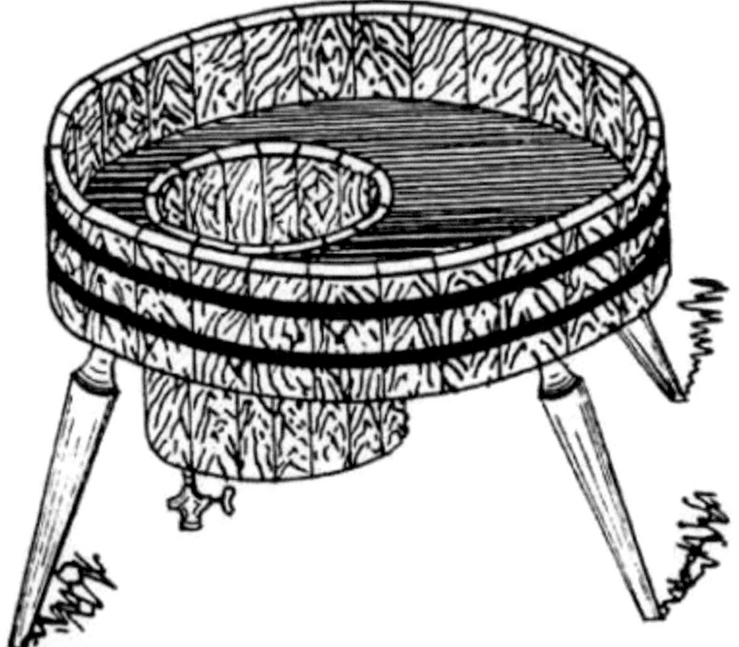

FIG. 27.—Dye-tub for Paranitroaniline Red.

The beta-naphthol bath does not keep well and in time tends to grow brown, and when this occurs stains are invariably produced on the cotton. When the yarn or cloth has been prepared with the beta-naphthol, and dried, the developing should be immediately proceeded with, for it is found that by allowing the prepared cotton to lie about it becomes covered with brown stains, and when such stained cotton is passed through the developing bath stains and defective dyeing result. [Pg 192]

It has been found that by adding a little tartar emetic to the beta-naphthol bath this is largely if not entirely prevented, and the prepared cloth may be kept for a reasonable length of time before proceeding with the development without fear of stains being formed.

Various additions have been made from time to time to the naphthol bath. Some of these take the form of special preparations of the colour manufacturers, and are sold as naphthol D, naphthol X, red developer C, etc., sometimes gum tragacanth has been added, at others in place of Turkey-red oil there is used a soap made from castor oil with soda and ammonia, but such complicated baths do not yield any better results than the simple preparing liquor given above.

FIG. 28.—Padding Machine for Paranitroaniline Red.

Dyeing Paranitroaniline Red on Piece Goods.

The dyeing of this red on to piece goods only differs from that on yarn by reason of the difference in the form of material that is dealt with. [Pg 193]

1. *Preparing or Grounding.*—The same liquor may be used. This operation is best done on a padding machine, a sketch of which is

given in Fig. 26, showing the course of the cloth through the liquor. This is contained in the box of the machine, and this is kept full by a constant stream flowing in from a store vat placed beside the machine. After going through the liquor, the cloth passes between a pair of squeezing rollers which squeeze out the surplus liquor. Fig. 28 shows a view of a padding machine adapted for grounding paranitroaniline reds. After the padding, the cloth is dried by being sent over a set of drying cylinders, or through what is known as the hot flue.

2. *The Developing.* — After being dried, the pieces are sent through a padding machine charged with the developing liquor made as described above, after which the cloth is rinsed, then soaped, and then washed. Some dyers use a continuous machine for these operations, such as shown in Fig. 29.

While the developing bath used for piece goods may be the same as that used for yarns, some dyers prefer to use one made somewhat differently, thus 6¼ lb. paranitroaniline are mixed with 7 gallons boiling-water and 1½ gallons hydrochloric acid; when dissolved 16 gallons of cold water are added, then, after completely cooling, 3½ lb. sodium nitrite dissolved in 3 gallons cold water. After twenty minutes, when the diazotisation is complete, water is added to make the whole up to 40 gallons. The acetate liquor is made from 13¼ lb. acetate of soda in 13½ gallons of water.

Equal quantities of these two liquors are used in making the developing bath.

Of late years, under the names of Azophor red P N, Nitrazol C, Nitrosamine, etc., there has been offered to dyers preparations of diazotised paranitroaniline in the form of a powder or paste, readily soluble in water, that will keep in a [Pg 194]

184

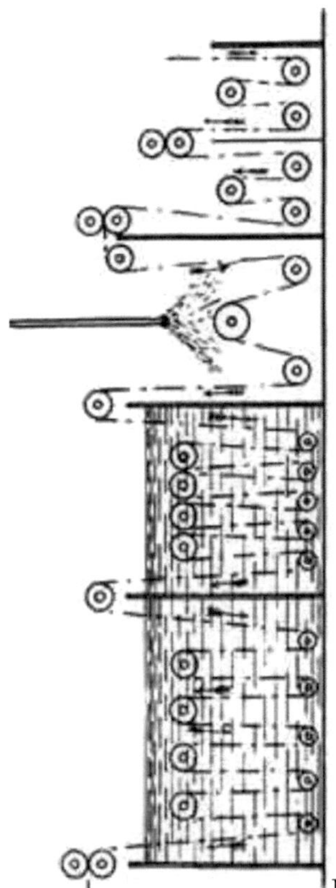

FIG. 29 — Developing Machine for Para-nitroaniline Red.

cool and dry place for any reasonable length of time. These are prepared in various ways, and to any dyer who does not [Pg 195] want the trouble of diazotising the paranitroaniline they offer some advantages. They produce a red equal in every respect to that obtained from paranitroaniline. The following details show the method to be followed with some of these products, others are very similar to make the developing baths.

Paranitroaniline Red with Nitrazol C. — Dissolve 25 lb. Nitrazol C in 12 gallons of cold water with constant stirring, then add sufficient cold water to make 37 gallons. In another vessel dissolve 11 lb. of acetate of soda in 5½ gallons water, then add 1½ gallons caustic soda, 36° Tw., mixed with 5½ gallons water. The developing bath is made by mixing both these solutions. It will suffice for both yarn and piece goods.

Paranitroaniline Red with Azophor Red P N.—Dissolve 5½ lb. of Azophor red P N in 4 gallons of water—it dissolves almost completely but usually a few particles of a flocculent character remain undissolved, these can be removed; 2¼ gallons of caustic soda lye of 36° Tw. are diluted with water to 10 gallons, and this is added with constant stirring to the azophor red P N solution. When all is mixed and a clear solution obtained, the developing bath is ready for use, and is used in the same way as the paranitroaniline bath.

Metanitroaniline Orange.—This orange is produced in the same way as the paranitroaniline red, using metanitroaniline or Azophor orange M N in place of the paranitroaniline or the Azophor red P N given for the red. The quantities of all the materials used are identical.

Nitrosamine Red.—Dissolve 5 lb. Nitrosamine red in 5 gallons of water and 2¼ lb. hydrochloric acid, when well mixed there is added 2½ lb. acetate of soda, when all is dissolved add sufficient water to make 6½ gallons. This bath is used exactly in the same way as the paranitroaniline developing bath, and it produces identical results in every way. [Pg 196]

Paranitroaniline Brown.—By boiling the paranitroaniline red dyed cotton in a weak bath of copper sulphate a very fine fast brown resembling a cutch brown is produced. A better plan, however, is to prepare the cotton with a ground containing an alkaline solution of copper, 3 lb. beta-naphthol are dissolved in 5 pints of caustic soda lye of 36° Tw., to which is added 5 lb. Turkey-red oil and 10 pints alkaline copper solution, water being added to make 13 gallons of liquor. The cotton is treated in this way as with the ordinary beta-naphthol preparation. The alkaline copper solution is made by taking 5 pints of copper chloride solution at 76° Tw., adding 3¼ lb. tartaric acid, 6 pints caustic soda lye, 70° Tw., and 2 pints of glycerine. The developing bath for the brown is the same as for the paranitroaniline red, or the Azophor red P N bath may be used.

Toluidine Orange.—For this colour the cotton is prepared with the beta-naphthol in the ordinary way. The developing bath is made from 2 lb. orthonitrotoluidine mixed with 12 pints boiling water and 2¼ pints hydrochloric acid; when dissolved allow to cool and then add 12½ lb. ice. When thoroughly cold stir in 2½ pints of sodium nitrite solution containing 3 lb. per gallon. Stir well for twenty minutes, then filter; add 4 lb. sodium acetate and sufficient ice-cold water to make 13 gallons. Use this bath in the same way as the paranitroaniline bath.

Beta-naphthylamine Red.—This red is a good one, but is not so bright or so fast as the paranitroaniline red, hence although somewhat older in point of time it is not dyed to the same extent. The developing bath is made from 1¾ lb. beta-naphthylamine dissolved with the aid of 10 pints boiling water and 1 pint hydrochloric acid. When dissolved allow to cool; add 27 lb. ice and 2 pints hydrochloric acid. When cooled to 32° to 36° Tw., add 3 pints sodium nitrite solution (3 lb. per gallon) and 4 lb. sodium acetate, making up to [Pg 197] 13 gallons with water. This also is used in precisely the same way as the paranitroaniline red developing liquor.

Alpha-Naphthylamine Claret.—This is a very fine and fairly fast red, and next to the paranitroaniline red may be considered the most important of the naphthol colours. The developing bath is a little more difficult to make, owing to the fact that it is more difficult to get the alpha-naphthylamine into solution. The best way of proceeding is the following: Heat 1¾ lb. of alpha-naphthylamine in 10 pints of boiling water, agitating well until the base is very finely divided in the water, then 1¼ pints of hydrochloric acid is added, and the heat and stirring continued until the base is dissolved, then the mass is allowed to cool, 27 lb. of ice is added and 1½ pints of hydrochloric acid. When cooled down to 32° to 36° F., there is added 3 pints sodium nitrite solution (3 lb. per gallon), and after allowing the diazotisation to be completed, 4 lb. sodium acetate and sufficient water to make 13 gallons of liquor.

The bath is used in the same manner as the previous developing baths.

Dianisidine Blue.—Dianisidine develops with beta-naphthol, a violet blue, which is not very fast, but by the addition of some copper to the developing bath a very fine blue is got which has a fair degree of fastness. The developing bath is made as follows: Mix 10½ oz.

dianisidine with 7 oz. hydrochloric acid and 7½ pints of boiling water, when complete solution is obtained it is allowed to cool, then 20 lb. of ice is added. Next 1¾ pints of nitrite of soda solution, containing 1½ lb. per gal. and 2½ pints of cold water. Stir for thirty minutes, then add 1¼ pints copper chloride solution at 72° Tw., and sufficient water to make up 6½ gallons.

The cotton is prepared with beta-naphthol in the usual way, and then passed through this developing bath. [Pg 198]

Amidoazotoluol Garnet.—Amidoazotoluol produces with beta-naphthol a fine garnet red in the usual way.

The developing bath is made from 14 oz. amidoazotoluol, mixed with 1½ pints of sodium nitrite solution containing 1½ lb. per gallon, when well mixed add 1 pint of hydrochloric acid diluted with 2 pints water, when this is well mixed add sufficient water to make up a gallon, then add 1 lb. acetate of soda.

The cotton is passed through this dye-bath, then washed well, passed through a weak acid bath, then soaped well, washed and dried.

(8) DYEING COTTON BY IMPREGNATION WITH DYE-STUFF SOLUTION.

Indigo is a dye-stuff which requires special processes for its application to the cotton or wool fibre.

Its peculiarity is that in the form in which it comes to the dyer it is insoluble in water, and to enable it to be dissolved and therefore to be used as a dye, the indigo has to go under a special treatment. The colouring principle of indigo is a body named indigotin, to which the formula $C_{16}H_{16}N_2O_2$ has been given. When indigo is mixed with substances like lime and copperas, lime and zinc, zinc and bisulphite of soda, which cause the evolution of nascent hydrogen, it takes up this body and passes into another substance which is called indigo white that has the formula $C_{16}H_{12}N_2O_2$, leuco, or white indigo; this substance is soluble in water, and so when it is formed the indigo passes into solution and can then be used for dyeing. But indigo white is an unstable substance on exposure to air, the oxygen of the latter attacks the hydrogen which it has taken up, and indigotin is reformed, the indigo white changing again into indigo blue.

Indigo dyeing consists of three operations: [Pg 199] —

(1) Preparation of the indigo solution, or, as it is called, setting the dye vat. (2) Steeping the cotton in this vat. (3) Exposing to the air.

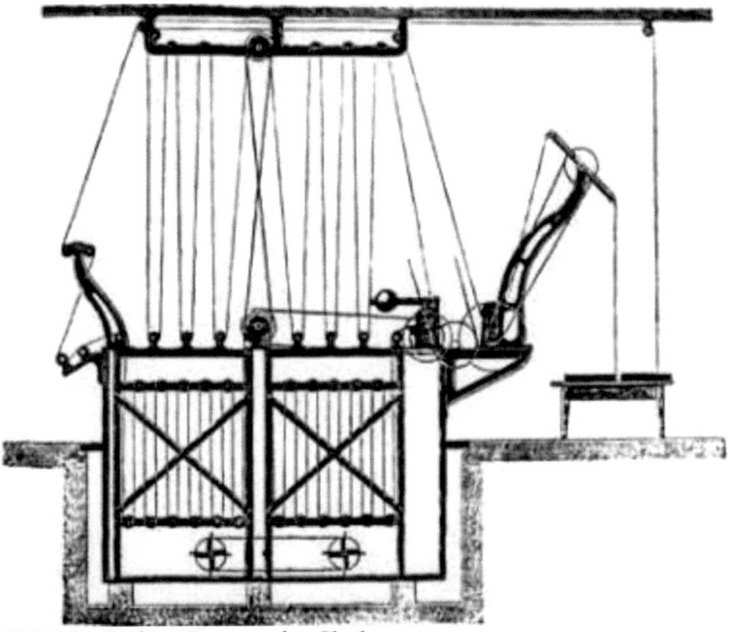

FIG. 30. — Indigo Dye-vat for Cloth.

There are several methods of preparing, or setting the dye vat, and of each of these modifications are in use in every indigo dye-house: —

(1) With lime and copperas. (2) Zinc and lime vat. (3) Zinc and bisulphite of soda.

In all cases it is necessary for the indigo to be ground to the form of a fine paste with water; this is usually done in what is known as the ball-grinding mill. The finer it is ground the more easy is it to make the dye-vats. [Pg 200]

The dye-vats may be either round tubs or square wooden tanks; when cloths or warps are being dyed these may be fitted with winces and guide rollers, so as to draw the materials through the liquor. In the case of yarns in hanks these appliances are not necessary.

Fig. 30 is a sketch of an indigo dye-vat for cloth or warps.

(1) **Lime and Copperas Vat.**—To prepare this vat take 75 gallons of water, 4 lb. of indigo, 8 lb. copperas, and 10 lb. of good quicklime. Put these into the vat in the order shown. The amount of indigo is added in proportion to the shade which is required to be dyed: for pale shades, 2 lb. to 3 lb. will be sufficient; while for deep shades, 6 lb. to 7 lb. may be used. The amount of copperas should be from one and a half to twice that of the indigo. The vat should be stirred very well and then left to stand. The changes which occur are probably the following: The lime acts upon the copperas and produces ferrous hydrate—this is unstable and tends to take up oxygen and hydrogen from the water, particularly when there is some indigo present, and forms ferric hydrate; hydrogen is at the same time liberated, and combines with the indigo to form the soluble indigo white. It takes about twenty-four hours to make an indigo vat. When properly made and in good condition, the liquor will be clear and of a brownish-yellow colour, a bluish scum may collect on the surface. If the liquor appears at all greenish it is an indication that the indigo has not been completely reduced, and the vat needs a further addition of lime and copperas, which should be of good quality. Too much of each should not be used, because with them there is formed at the bottom of the vat a sediment of calcium sulphate and ferric hydrate, and it is not wise to increase this to too great an extent, which would be the case if too much lime and copperas were added.

To Use this Vat.—Any scum on the surface is raked on one [Pg 201] side, the cotton yarn immersed for a few minutes, then it is taken out, wrung, allowing the excess liquor to flow back into the vat, and the yarn hung up in the air for the blue to develop. The depth of shade which is dyed depends chiefly upon the amount of indigo in the vat, and also upon the time during which the hanks are dipped in the liquor. Light and medium shades can be readily and conveniently got by a single dip, but deep shades are best got by repeating the dipping once or twice as occasion demands. Deep shades got by using a strong bath at a single dip are found to rub badly, while by repeated dips the dye gets more into the substance of the fibre, and therefore the colour is more firmly fixed and it rubs less.

Some indigo dyers have quite a range of vats, using those fresh made for dyeing deep shades, while the old vats being nearly exhausted are used only for light shades and finally when completely exhausted are thrown away. After the day's work the vat should be stirred up and then allowed to stand. If necessary it may be strengthened by the addition of fresh quantities of indigo, lime and copperas, the next morning it will be ready for use. Generally a lime-copperas vat will remain in good working order for about a month, when it will be necessary to throw it away.

(2) Zinc and Lime Vat.—Zinc dust is a bye-product in the process of zinc extraction. It is a grey, very heavy powder, consisting mostly of finely divided metallic zinc, with traces of oxide and sulphide of zinc. Of these only the metallic zinc is active in reducing the indigo, the rest of the ingredients are not of any consequence. The valuation of zinc dust is a very difficult operation, but it is desirable that this be done, as the product is liable to be very variable in the proportion of actual zinc it contains, and it will pay large buyers always to have it tested. Zinc dust must always be kept in a dry place. [Pg 202]

For the reduction of zinc powder lime is chiefly used. The following are two good mixtures.

Vat with zinc and lime:—

10 lb. indigo, dry and ground fine; 5½ lb. zinc dust; 22 lb. slaked lime, dry.

The vat is set as follows, a part of the lime is mixed with the indigo, and the two bodies are well mixed together and allowed to stand for ten minutes, then the zinc powder is added. It is best to make this into a smooth paste with water before adding it to the other ingredients, then the rest of the lime is added and the whole is thoroughly stirred together with the necessary quantity of water.

Vat with zinc powder, lime and soda:—

10 lb. indigo, dry and ground fine; 10 lb. zinc powder; 10 lb. slaked lime, dry; 35 lb. caustic soda at 11° Tw.

Add the lime to the ground indigo, then add the zinc and finally the soda lye.

Soon after the various ingredients of the vats are added together the whole mass becomes hot, when it must be well stirred. It soon begins to evolve gas and the mixture froths. In from two to four hours the evolution of gas ceases. The dark blue solution now becomes yellow and the liquor shows all the characteristics of the indigo vat. It is necessary to keep the vat well stirred up during the time of setting, which takes from five to six hours. If there is much evolution of gas after this time it indicates that too much zinc powder has been added; this is a common fault with dyers, and such excess causes the vat to be too much disturbed and to work dirty. A lime-zinc vat, with occasional additions of new [Pg 203] materials, keeps good for three months, and even then is in a better condition than the copperas vat.

This vat is used in precisely the same way as the copperas vat; as it contains no sediment, or but little, it works cleaner than the copperas vat and as a rule the indigo blues dyed in it are faster to rubbing.

After a day's work it can be well stirred up and fresh additions of lime, zinc and indigo made to bring it up to its original dyeing strength.

(3) Zinc-Bisulphite Indigo Vat.—When zinc dust and bisulphite of soda are mixed together a reaction sets in, the zinc dissolves, and there is formed sodium hydrosulphite and zinc and sodium sulphites. If now indigo is mixed with this solution the sodium hydrosulphite exerts a reducing action on it, forming white indigo and sodium sulphite, a perfectly clear solution being obtained, which may be used in dyeing cotton or wool.

With this vat it is customary to prepare a strong stock solution of reduced indigo, and to add this to the dyeing vats as may be required.

To Make the Stock Liquor.—Take 20 lb. of indigo, grind into a paste with 20 gallons of boiled water, then add 25 lb. lime slaked into a milk. In a separate tub there is mixed 80 lb. bisulphite of soda, 70° Tw., with 9 lb. zinc dust; this mixture is well stirred and every care taken to prevent it getting hot. When the zinc has dissolved and the mixture is free from any sulphurous smell it is run into the indigo mixture given above. The whole is well stirred together for some time, and then at intervals, until the indigo has become dissolved, sufficient water is added to make up 50 gallons, when the stock liquor will be ready. It should have a deep yellow colour. The surface may have a scum of a bronzy colour collect on it. This stock liquor should be kept in casks free from exposure to the air. [Pg 204]

To make the working vat from this stock liquor the following is the method of proceeding: —

Water is run into the vat, and this is heated from 70° to 80° C. in order to expel air from it, after which it may be allowed to cool, then for each 1,000 gallons contained in the vat there is added 30 lb. bisulphite of soda, 3 lb. zinc dust and 3 lb. lime, made into a cream. When all these ingredients are dissolved a quantity of the stock liquor is added in proportion to the shade that it is desired to dye. The whole is well stirred, then the vat is allowed to rest for half an hour to enable any sediment to settle, and then the dyeing is proceeded with.

Should the vat show signs of becoming green in colour it is a good plan to add a mixture of 1 lb. zinc dust and 10 lb. bisulphite of soda. The vat should be kept alkaline, and so a little lime may be added from time to time.

After a day's work it is well to add a little of the zinc and bisulphite mixture, to stir well and allow to stand overnight; the next morning strengthen up the vat by adding fresh stock liquor.

In place of using lime in making up the vat it is possible to use a mixture of caustic soda and ammonia. The lime will tend to cause some sediment to form in the vat, whereas the soda and ammonia will not. When they are used the following mode of working may be followed: —

Stock Liquor. —Soda zinc vat: Put in a tub 26 gallons cold water, 15 lb. zinc powder ground into a paste with 6 gallons water, then stir in 8 gallons bisulphite of soda at 60° Tw., stir well, keeping the heat down as much as possible, after which add 8 pints caustic soda lye at 70° Tw. and 14 pints 20 per cent. liquor ammonia. When all is thoroughly mixed add 30 lb. indigo, ground into a paste with 7 gallons water, allow to stand for half an hour, then add water to make [Pg 205] 100 gallons, stir at intervals for twelve hours or so, when the stock liquor will be ready for use.

This is used to make the vat in the same way as the first above described. It is needful before adding the stock liquor in making a vat to destroy any oxygen or air which is present in the vat. This may most conveniently be done by adding thereto a solution of hydrosulphite of soda, which may be made by mixing 4½ lb. zinc dust with 5 gallons of water and 3 gallons bisulphite of soda at 70° Tw., stirring well, so that the temperature does not rise too high, there is then added 2 pints caustic soda and 3 pints liquor ammonia, 20 per cent.; when all is dissolved, water is added to make up 13 gallons.

Should the vat show signs of becoming charged with indigo, indicated by its becoming of a green colour, a little of this hydrosulphite added from time to time will correct it.

The hydrosulphite-indigo vat made by either of the two methods indicated above works well, and with due care may be kept in work for months. It gives good shades of indigo, although some dyers consider that these have not the rich bronze hue got from the lime and copperas vat. The shades are rather faster to rubbing.

It has been proposed to employ the dye-stuff indophenol in conjunction with indigo, in which case the method of making the vat is with zinc, bisulphite of soda, caustic soda and ammonia as last described, only in place of using all indigo a mixture of 22 lb. indigo and 7½ lb. indophenol is used. Good blue shades of considerable fastness can thus be got.

Aniline Black. — This black is produced direct upon the cotton fibre by various processes which entail the oxidation of aniline. The chemical composition and constitution of aniline black has not yet been worked out. It is not by any means an easy colour to dye, but still with careful attention [Pg 206] to carrying out the various operations in detail excellent results can be attained.

Aniline black is the fastest black which is known, it resists when well dyed exposure to air and light, is quite fast to washing and soaping. Its disadvantages are that there is, with some methods of working, a tendency to tender the cotton fibre, making it tear easily; secondly, on exposure to air it tends to turn green, this however only happens when the black has not been dyed properly.

At the present day it is perhaps without doubt the most used of all blacks. The methods for producing it are many and varied, the following recipes show how some of the finest aniline blacks can be dyed: —

Ungreenable black is formed when the aniline is by the action of the oxidising agents converted into a substance named nigraniline. This compound when formed will not turn green on the fibre under the influence of acids.

1. The most usual oxidising agent employed for dyeing aniline black is bichromate of soda, which salt will be found much better for all purposes than bichromate of potash. Two separate solutions are prepared: (1) 61 lb. aniline, 9 lb. hydrochloric acid and 10 gallons of water; and (2) 12 lb. bichromate of soda and 20 gallons of water. After cooling, equal quantities of these solutions are mixed and the cotton worked rapidly through the mixture, in a few minutes it assumes a bronze black. The material is then wrung out and steamed for twenty minutes at 3½ lb. pressure, which process renders it jet black and also ungreenable.

2. Another aniline black: For 100 lb. cotton use 11 lb. aniline oil, 15 lb. bichromate of soda, 40 lb. hydrochloric acid and 160 gallons water or 12 lb. sulphuric acid. The dye-bath is filled with the water and the cold solution of aniline oil and a part of the hydrochloric acid in water is first added, afterwards the bichromate is dissolved in a small quantity [Pg 207] of water, working cold at first and gradually rising to the boil.

3. Another method is the following and gives a black that is fast and ungreenable and will not rub: 10 lb. chlorate of soda, 10 lb. ammonium chloride, 10 lb. copper sulphate, 35 lb. aniline salt, 101 lb. aniline oil and 20 gallons water. The sodium chlorate and ammonium chloride are dissolved in 6½ gallons and the copper sulphate separately in 5½ gallons water. The aniline salt is dissolved in

as little hot water as possible and neutralised with a small amount of aniline oil (10 lb.). The solution of aniline salt is first added to the bath, then the sodium chlorate and ammonium chloride, and lastly the copper sulphate, dilute the whole to 14° Tw. and then enter the goods. Next steam, then run through a solution containing 10 lb. bichromate and 5 lb. soda per 100 gallons water at 160° F., after which the goods are washed and dry steamed at 15 lb. pressure.

4. A very good black is the prussiate or steam aniline black whose cheapness should recommend it. Prepare concentrated solutions of 1¾ lb. aniline salt in 1 gallon water, 1½ lb. ferrocyanide of potash in ¾ gallons water and 1½ lb. potassium chlorate in 1½ gallons water. Mix the solutions and work in a jigger, then steam in a Mather & Platt apparatus for two minutes, then work hot in a jigger in a solution of 2 lb. bichromate per 50 gallons water, dry and finish.

Either of these methods yields a good full black; with a little experience and care perfectly uniform shades will be got. [Pg 208]

CHAPTER V.

DYEING UNION (MIXED COTTON AND WOOL) FABRICS.

There is now produced a great variety of textile fabrics of every conceivable texture by combining the two fibres, cotton and wool, in a number of ways; the variety of these fabrics has of late years considerably increased, which increase may be largely ascribed to the introduction of the direct dyeing colouring matters—the Diamine dyes, the Benzo dyes, the Congo and the Zambesi dyes, for in the dyeing of wool-cotton fabrics they have made a revolution. The dyer of union fabrics, that is, fabrics composed of wool and cotton, was formerly put to great straits to obtain uniform shades on the fabrics supplied to him, owing to the difference in the affinity of the two fibres for the dye-stuffs then known. Now the direct dyes afford him a means of easily dyeing a piece of cotton-wool cloth in any colour of a uniform shade, while the production of two coloured effects is much more under his control, and has led to the increased production of figured-dress fabrics, with the ground in one fibre (wool) and colour, and the design in another fibre (cotton) and colour. The number of direct dyes issued by the various colour

196

manufacturers is so great that it would take a fairly considerable space to discuss them all.

To obtain good results it is needful that the dyer of union fabrics should have a thorough knowledge of the dyes he is using, for each dye makes a rule to itself as regards its power of dyeing wool and cotton—some go better [Pg 209] on to the cotton than on to the wool, and *vice versa*. Some dye wool best at the boil, others equally well below that heat; some go on the cotton at a moderate temperature, others require the dye-bath to be boiling; some will go on to the cotton only, and appear to ignore the wool.

The presence or absence in the dye-bath of such bodies as carbonate of soda, Glauber's salt, etc., has a material influence on the degree of the affinity of the dye-stuff for the two fibres, as will perhaps be noted hereafter. Again, while some of the dyes produce equal colours on both fibres, there are others where the tone is different. With all these peculiarities of the Diamine and other direct dyes the union dyer must make himself familiar. These dyes are used in neutral baths, that is, along with the dye-stuff. It is often convenient to use, along with the direct dyes, some azo or acid dyes, which have the property of dyeing the wool from neutral baths, many examples of such will be found in the practical recipes given below. The dyes now under consideration may be conveniently classed into five groups.

1. Those dyes which dye the cotton and wool from the same bath to the same shade, or nearly so. Among such are Thioflavine S, Diamine fast yellow B, Diamine orange B, Diamine rose B D, Diamine reds 4 B, 5 B, 6 B and 10 B, Diamine fast red F, Diamine Bordeaux B, Diamine brown N, Diamine browns 3 G, B and G, Diamine blues R W, B X, Diamine blue G, Diamine greens G and B, Diamine black H W, Diamine dark blue B, Union blacks B and S, Oxydiamine blacks B, M, D and A, Diamine catechine G, Union blue B B, Oxyphenine, Chloramine yellow, Alkali yellow R, Chromine G, Titan scarlet S, Mimosa, Curcumine, Primuline, Auroline, Congo Corinth B, Thiazole yellow, Columbia yellow, Oxydiamine yellow G G, Oxydiamine oranges G and R, Diamine orange F, Oxydiamine red S.

2. Dyes which dye the cotton a deeper shade than the [Pg 210] wool. The following belong to this group: Diamine fast yellow A,

Diamine oranges G and D, Diamine catechine G, Diamine catechine B, Diamine sky blue, Diamine blue 2 B, Diamine blue 8 B, Diamine blue B G, Diamine brilliant blue G, Diamine new blue R, Diamine steel blue L, Diamine black R O, Diamine black B 0, Diamine black B H, and Oxydiamine black S O O O, Diamine nitrazol brown G, Diamine sky blue F F, Diamine dark blue B, Diamine Bordeaux B, Diamine violet N, Oxydiamine violet B, Columbia blacks B and F B, Zambesi black B, Congo brown G, Direct yellow G, Direct orange B, Clayton yellow, Cotton yellow, orange T A, Benzo purpurine B, Brilliant Congo R, Chicago blues B and 4 B and 6 B.

3. Dyes which dye wool a deeper shade than the cotton. The dyes in this group are not numerous. They are Diamine gold, Diamine scarlet B, Diamine scarlet 3 B, Diamine Bordeaux S, Diamine blue R W, and Diamine green G, Diamine reds N 0 and B, Chicago blues G and R, Brilliant purpurine R, Diamine scarlet B, Delta purpurine 5 B, Chrysamine, Titan blue, Titan pink, Congo oranges G and R, Erie blue 2 G, Congo R, Brilliant Congo R, Erika B N, Benzo purpurines 4 B and 10 B, Chrysophenine, Titan yellow, Titan browns Y, R and O, Congo brown G, Sulphon azurine B, Zambesi black B.

4. Dyes which produce different shades on the two fibres. Diamine brown G, and Diamine blue 3 R, Diamine brown V, Diamine brown S, Diamine nitrazol brown B, Diamine blues B X and 3 R, Diamine blue black E, Benzo blue black G, Benzo purpurine 10 B, Benzo azurines R, G and 3 G, Columbia red S, Brilliant azurine 5 G, Titan marine blue, Congo Corinths G and B, Azo blue, Hessian violet, Titan blue, Azo mauve, Congo brown, Diamine bronze G, Zambesi browns G and 2 G, Zambesi black F.

[Pg 211]

5. Azo-acid dyes, which dye wool from neutral baths, and are therefore suitable for shading up the wool to the cotton in union fabric dyeing. Among the dyes thus available may be enumerated: Naphthol blues G and R, Naphthol blue black, Formyl violet 10 B, Lanacyl blue B B, Lanacyl blue R, Alkaline blue, Formyl violets S 4 B and 6 B, Rocceleine, Azo red A, Croceine A Z, Brilliant scarlet, Orange extra, Orange E N Z, Indian yellow G, Indian yellow R, Tropæoline O O, Naphthylamine black 4 B and Naphthol blue black, Brilliant scarlet G, Lanacyl violet B, Brilliant milling green B,

198

Thiocarmine R, Formyl blue B, Naphthylamine blacks D, 4 B and 6 B; Azo-acid yellow, Curcumine extra, Mandarine G, Ponceau 3 R B, Acid violet 6 B, Guinea violet 4 B, Guinea green B, Wool black 6 B.

Regarding the best methods of dyeing, that in neutral baths yields the most satisfactory results in practical working. It is done in a boiling hot or in a slightly boiling bath, with the addition of 6¼ oz. crystallised Glauber's salt per gallon water for the first bath, and when the baths are kept standing 20 per cent. crystallised Glauber's salt, reckoned upon the weight of the goods, for each succeeding lot.

In dyeing unions, the dye-baths must be as concentrated as possible, and must not contain more than from 25 to 30 times as much water as the goods weigh. In this respect it may serve as a guide that concentrated baths are best used when dyeing dark shades, while light shades can be dyed in more diluted baths. The most important factor for producing uniform dyeings is the appropriate regulation of the temperature of the dye-bath. Concerning this, the dyer must bear in mind that the direct colours possess a greater affinity for the cotton if dyed below the boiling point, and only go on the wool when the bath is boiling, especially so the longer and more intensely the goods are boiled.

The following method of dyeing is perhaps the best one: Charge the dye-bath with the requisite dye-stuff and Glauber's [Pg 212] salt, boil up, shut off the steam, enter the goods and let run for half an hour without steam, then sample. If the shade of both cotton and wool is too light add some more of the dye-stuffs used for both fibres, boil up once more and boil for a quarter to half an hour. If the wool only is too light, or its shade different from that of the cotton, add some more of the dye-stuff used for shading the wool and bring them again to the boil. If, however, the cotton turns out too light, or does not correspond in shade to the wool, add some more of the dye-stuffs used for dyeing the cotton, without, however, raising the temperature. Prolonged boiling is only necessary very rarely, and generally only if the goods to be dyed are difficult to penetrate, or contain qualities of wool which only with difficulty take up the dye-stuff. In such cases, in making up the bath dye-stuffs are to be selected some of which go only on the wool and others which go only on the cotton (those belonging to the second group).

The goods can then be boiled for some time, and perfect penetration and level shades will result. If the wool takes up the dye-stuff easily (as is frequently the case with goods manufactured from shoddy), and are therefore dyed too dark a shade, then dye-stuffs have to be used which principally dye the cotton, and a too high temperature should be avoided. In such cases it is advisable to diminish the affinity of the wool by the addition of one-fifth of the original quantity of Glauber's salt (about 3/8 oz. per gallon water), and from three-quarters to four-fifths of the dye-stuff used for the first lot. Care has to be taken that not much of the dye liquor is lost when taking out the dyed goods, otherwise the quantities of Glauber's salt and dye-stuff will have to be increased proportionately. Wooden vats, such as are generally used for piece dyeing, have proved the most suitable. They are heated with direct, or, still better, with indirect steam. The method which has proved most advantageous is to let [Pg 213] the steam run into a space separated from the vat by a perforated wall, into which space the required dye-stuffs and salt are placed.

The mode of working is rather influenced by the character of the goods, and the following notes will be found useful by the union dyer:—

Very little difficulty will be met with in dyeing such light fabrics as Italians, cashmere, serges and similar thin textiles lightly woven from cotton warp and woollen weft. When deep shades (blacks, dark blues, browns and greens), are being dyed it is not advisable to make up the dye-bath with the whole of the dyes at once. It is much better to add these in quantities of about one-fourth at a time at intervals during the dyeing of the piece. It is found that the affinity of the wool for the dyes at the boil is so much greater than is that of the cotton that it would, if the whole of the dye were used, take up too much of the colour, and then would come up too deep in shade. Never give a strong boil with such fabrics, but keep the bath just under the boil, which results in the wool dyeing much more nearly like to cotton.

Bright Yellow.—Use 2 lb. Thioflavine S in a bath which contains 4 lb. Glauber's salt per 10 gallons of dye liquor.

Good Yellow.—A very fine deep shade is dyed with 2½ lb. Diamine gold and 2½ lb. Diamine fast yellow A, in the same way as the last. Here advantage is taken of the fact that while the Diamine gold dyes the wool better than the cotton, the yellow dyes the cotton the deeper shade, and between the two a uniform shade of yellow is got.

Pale Gold Yellow.—Use a dye liquor containing 4 lb. Glauber's salt in every 10 gallons, 2½ lb. Diamine fast yellow A, 2 oz. Indian yellow G and 3½ oz. Indian yellow R. In this recipe there is used in the two last dyes purely wool yellows, which dye the wool the same tint as the fast yellow A dyes the cotton. [Pg 214]

Bright Yellow.—Use in the same way as the last, 2½ lb. Diamine fast yellow B and 3 oz. Indian yellow G.

Gold Orange.—Use as above 2 lb. Diamine orange G, 5½ oz. Indian yellow K and 1½ oz. Orange E N Z.

Deep Orange.—Use 2½ lb. Diamine orange D C, 6½ oz. Orange E N Z, and 3¼ oz. Indian yellow R.

Black.—Use 4½ lb. Union black S, 2 oz. Diamine fast yellow A, 5 oz. Naphthol blue black and 3¼ oz. Formyl violet S 4 B, with 4 lb. Glauber's salt in each 10 gallons dye liquor.

Navy Blue.—Use 1¼ lb. Union black S, 3 lb. Diamine black B H, ½ oz. Naphthol blue black, ½ lb. Formyl violet S 4 B and 2½ oz. alkaline blue B.

Red Plum.—Use a dye-bath containing 2½ lb. Oxydiamine violet B and 3¼ oz. Formyl violet S 4 B.

Dark Green.—A fine shade can be dyed in a bath containing 3 lb. Diamine green B and 1½ lb. Diamine black H W.

Dark Slate.—Use 4 lb. Diamine black H W, 2 oz. Naphthol blue black and 3 oz. Azo red A.

Sage.—Use a dye-bath containing 4 lb. Diamine bronze G and 1¼ oz. Naphthol blue black.

Dark Brown.—A fine dark shade is got from 2½ lb. Diamine brown V and 2 oz. Naphthol blue black.

Peacock Green. — Use 3¾ lb. Diamine steel blue L, 13 oz. Diamine fast yellow B, 14½ oz. Thiocarmine K and 2¼ oz. Indian yellow G in a bath of 4 lb. Glauber's salt per gallon dye liquor.

Dark Sea Green. — Use 9 oz. Diamine steel blue L, 3¾ oz. Diamine fast yellow B, ½ oz. Diamine orange G, 1¼ oz. Naphthol blue black and ¾ oz. Indian yellow G.

Dark Brown. — Use 1 lb. Diamine orange B, 1 lb. Diamine Fast yellow S, 13¾ oz. Union black S, 1 lb. Diamine brown M and ½ lb. Indian yellow G. Fix in an alum bath after dyeing.

Dark Stone. — Use ½ lb. Diamine orange B, 3¾ oz. Union [Pg 215] black, ¼ oz. Diamine Bordeaux B, 1½ oz. Azo red A and ¾ oz. Naphthol blue black.

Black. — A very fine black can be got from 3½ lb. Oxydiamine black B M, 2 lb. Union black S, 9½ oz. Naphthol blue black and 4 oz. Formyl violet S 4 B.

Dark Grey. — A fine bluish shade of grey is got from 7 oz. Diamine black B H, 2¼ oz. Diamine orange G, 2½ oz. Diamine orange G, 2½ oz. Naphthol blue black and 1 oz. Orange E N Z.

Dark Blue. — A fine shade is got by using 2 lb. Diamine black B H, ½ lb. Diamine black H W, and 3½ oz. Alkaline blue 6 B.

Drab. — Use 3½ oz. Diamine orange B, ¾ oz. Union black, 1/8 oz. Diamine Bordeaux B, ¾ oz. Azo red A and ¼ oz. Naphthol blue black.

Plum. — Use 2½ lb. Diamine violet N, 9½ oz. Union black and 1 lb. Formyl violet S 4 B.

Bright Yellow. — Use a dye-bath containing 4 lb. Thioflavine S, 2 lb. Naphthol yellow S, 10 lb. Glauber's salt and 2 lb. acetic acid.

Pink. — Use 1/6 oz. Diamine Rose B D, ¼ oz. Diamine scarlet B, ½ oz. Rhodamine B and 20 lb. Glauber's salt.

Scarlet. — A fine shade is got from 1½ lb. Diamine scarlet B, ½ oz. Diamine red 5 B and 20 lb. Glauber's salt.

Orange. — Use a dye-bath containing 3½ lb. Diamine orange G, 14½ oz. Tropæoline O O, and 2¾ oz. Orange extra.

Sky Blue. — Use 1½ oz. Diamine sky blue and 1¼ oz. Alkaline blue B.

Bright Blue. — A fine shade similar to that formerly known as royal blue is got by using 1½ lb. Diamine brilliant blue G and 9¼ oz. Alkaline blue 6 B.

Maroon. — Use 3 lb. Diamine Bordeaux B, 2 lb. Diamine violet N and 3¼ oz. Formyl violet S 4 B.

Green. — A fine green similar in shade to that used for billiard-table cloth is got from 2 lb. Diamine fast yellow B, 2 [Pg 216] lb. Diamine steel blue L, 14½ oz. Thiocarmine R and 7¼ oz. Indian yellow G.

Gold Brown — A fine brown is got from 3 lb. Diamine orange B, ½ lb. Union black, 2½ oz. Diamine brown, ¾ oz. Naphthol blue black and ½ lb. Indian yellow G.

Navy Blue. — Use 3¼ lb. Diamine black B H, 1½ lb. Diamine brilliant blue G and ½ lb. Alkaline blue.

Fawn Drab. — A fine shade is got by dyeing in a bath containing 6¾ oz. Diamine orange B, 1¾ lb. Union black, ¼ oz. Naphthol blue black, ¼ oz. Diamine Bordeaux B and 1 oz. Azo red A.

In all these colours the dye-baths contain Glauber's salt at the rate of 4 lb. per 10 gallons.

Dark Brown. — 2½ lb. Diamine orange B, 13 oz. Diamine Bordeaux B, 1½ lb. Diamine fast yellow B, 1¾ lb. Union black and 3½ oz. Naphthol black.

Drab. — 1¾ lb. Diamine fast yellow R, 3¼ oz. Diamine Bordeaux B, 2½ oz. Union black, ½ oz. Naphthol blue black and 1¼ oz. Indian yellow G.

Dark Blue. — Use in the dye-bath 4¼ lb. Diamine dark blue B, 1½ lb. Diamine brilliant blue G, ¾ lb. Formyl violet S 4 B and 5 oz. Naphthol blue black.

Blue Black- — Use 3½ lb. Union black S, 1½ lb. Oxydiamine black B M, 6½ oz. Naphthol blue black and ¼ lb. Formyl violet S 4 B.

Dark Walnut. — 2¾ lb. Diamine brown M, 1½ lb. Union black S, and 11¼ oz. Indian yellow G.

Peacock Green.—Use in the dye-bath 3 lb. Diamine black H W, 5-1/6 oz. Diamine fast yellow B, 1¼ lb. Thiocarmine R and 1-1/6 oz. Indian yellow G.

Slate Blue.—Use in the dye-bath 6½ oz. Diamine carechine B, 4¾ oz. Diamine orange B, 2½ oz. Union black, 2¾ oz. Orange E N Z, and 1¾ oz. Naphthol blue black.

Dark Sage.—A good shade is dyed with 1 lb. Diamine [Pg 217] orange B, 6½ oz. Union black, 1¾ oz. Diamine brown M, 3¼ oz. azo red A and 2¼ oz. Naphthol blue black.

Navy Blue.—Use 2 lb. Diamine dark blue B, 1¼ lb. Lanacyl violet B, and 7 oz. Naphthol blue black.

Bronze Green.—A good shade is dyed with 2 lb. Diamine orange B, 5 oz. Diamine brown N, ¾ lb. Union black S, 1 lb. Indian yellow G and 2½ oz. Naphthol blue black.

Black.—Use 2½ lb. Oxydiamine black B M and 1½> lb. Naphthylamine black 6 B. Another recipe, 2¼ lb. Oxydiamine black B M, 1 lb. Diamine brown M, 1 lb. Orange E N Z and 2 oz. Naphthol blue black.

Dark Brown.—Use 1½ lb. Oxydiamine black B M, 15½ oz. Diamine brown M, 1¾ lb. Indian yellow G and 2¾ oz. Naphthol blue black. Another combination, 1½ lb. Oxydiamine black B M, 1½ lb. Orange E N Z, 1 lb. Indian yellow G and 5 oz. Naphthol blue black.

Scarlet.—3 lb. Benzo purpurine 4 B, ¾ oz. Ponceau 3 R B and ½ lb. Curcumine S.

Crimson.—½ lb. Congo Corinth G, 2 lb. Benzo purpurine 10 B and ½ lb. Curcumine S.

Bright Blue.—2 lb. Chicago blue 6 B, 3 oz. Alkali blue 6 B, 1½ oz. Zambesi blue R X. After dyeing rinse and develop in a bath of 8 oz. sulphuric acid in 10 gallons of water, then rinse well.

Dark Blue.—2½ lb. Columbia fast blue 2 G, 3 oz. Sulphon azurine D, 8 oz. Alkali blue 6 B. After dyeing rinse and develop in a bath of 8 oz. sulphuric acid in 20 gallons of water.

Orange.—9 oz. Congo brown G, 1½ lb. Mikado orange 4 R O and 1½ oz. Mandarine G.

Dark Green.—2 lb. Columbia green, ½ lb. Sulphon azurine D, 1 lb. Zambesi blue B X, 1½ oz. Curcumine S.

Black.—4 lb. Columbia black F B and 2 lb. Wool black 6 B.

Pale Sage Green.—5 oz. Zambesi black D, ¾ lb. Chrysophenine G and 1½ lb. Curcumine S. [Pg 218]

Slate.—½ lb. Zambesi black D, ¾ oz. Zambesi blue R X, ½ oz. Mikado orange 4 R 0 and 1½ oz. Acid violet 6 B.

Dark Grey.—1 lb. Columbia black F B, 3 oz. Zambesi black B and ¾ oz. Sulphon azurine D.

Drab.—1½ oz. Zambesi black D, ¾ oz. Mandarine G extra, ¼ oz. Curcumine extra and 3 oz. Mikado orange 4 R O.

Brown.—5 oz. Zambesi black D, ¾ oz. Mandarine G extra, 1½ oz. Orange T A and 2 oz. Mikado orange 4 R 0.

Nut Brown.—¾ lb. Congo brown G, ¼ lb. Chicago blue R W and ¾ lb. Mikado orange 4 R 0.

Dark Brown.—1 lb. Congo brown G, 1½ lb. Benzo purpurine 4 B, 1½ lb. Zambesi black F and ½ lb. Wool black 6 B.

Stone.—1 oz. Zambesi black D, ¼ oz. Mandarine G, ¼ oz. Curcumine extra and 1¼ oz. Mikado orange 4 R 0.

Slate Green.—3 oz. Zambesi black D, 1½ oz. Guinea green B.

Sage Brown.—½ lb. Zambesi black D, 1½ oz. Mandarine G extra, 3 oz. Curcumine extra, 3 oz. Acid violet 6 B, 6 oz. Mikado orange 4 R 0 and 4½ oz. Curcumine S.

Cornflower Blue.—3 oz. Chicago blue 4 R, ¼ lb. Zambesi blue R X, ¼ lb. Acid violet 6 B and ¾ oz. Zambesi brown G.

Dark Brown.—1½ lb. Brilliant orange G, ½ lb. Orange T A, 1 lb. Columbia black F B and ¼ lb. Wool black 6 B.

Dark Blue.—2 lb. Chicago blue W, 1 lb. Zambesi blue R X, ½ lb. Columbia black F B, 10 oz. Guinea green B and ½ lb. Guinea violet 4 B.

The Janus dyes may be used for the dyeing of half wool (union) fabrics. The best plan of working is to prepare a bath with 5 lb. of sulphate of zinc; in this the goods are worked at the boil for five minutes, then there is added the dyes previously dissolved in water, and the working continued for a quarter of an hour; there is then added 20 lb. Glauber's salt, and the working at the boil continued for one hour, at the end of which time the dye-bath will be fairly [Pg 219] well exhausted of colour. The goods are now taken out and put into a fixing-bath of sumac or tannin, in which they are treated for fifteen minutes; to this same bath there is next added tartar emetic and 1 lb. sulphuric acid, and the working continued for a quarter of an hour, then the bath is heated to 160° F., when the goods are lifted, rinsed and dried. In the recipes the quantities of the dyes, sumac or tannin and tartar emetic are given only, the other ingredients and processes are the same in all.

Dark Blue. — 2¼ lb. Janus dark blue B and ¼ lb. Janus green B in the dye-bath, and 16 lb. sumac extract and 2 lb. tartar emetic in the fixing-bath.

Blue Black. — 3½ lb. Janus black I and ½ lb. Janus black II in the dye-bath, and 16 lb. sumac extract and 2 lb. tartar emetic in the fixing-bath.

Dark Brown. — 2½ lb. Janus brown B, 1 lb. Janus black I, 3½ oz. Janus yellow G and 5 oz. Janus red B in the dye-bath, with 16 lb. sumac extract and 2 lb. tartar emetic in the fixing-bath.

Drab. — 1½ oz. Janus yellow R, ¾ oz. Janus red B, 1 oz. Janus blue R and ¼ oz. Janus grey B B in the dye-bath, and 4 lb. sumac extract and 1 lb. tartar emetic in the fixing-bath.

Grey. — 5 oz. Janus blue R, 3¼ oz. Janus grey B, 1½ oz. Janus yellow R and ¼ oz. Janus red B in the dye-bath, with 4 lb. sumach extract and 1 lb. tartar emetic in the fixing-bath.

Nut Brown. — 1 lb. Janus brown R, 8 oz. Janus yellow R and 1½ oz. Janus blue B in the dye-bath, and 8 lb. sumac extract and 1 lb. tartar emetic in the fixing-bath.

Walnut Brown. — 3 lb. Janus brown B, 1 lb. Janus red B, 1 lb. Janus yellow R, 1¼ oz. Janus green B in the dye-bath, with 8 lb. sumac extract and 1 lb. tartar emetic in the fixing-bath.

Crimson. — 2½ lb. Janus red B and 8 oz. Janus claret red B in the dye-bath, with 8 lb. sumac extract and 1 lb. tartar emetic in the fixing-bath. [Pg 220]

Dark Green. — 1½ lb. Janus green B, 1½ lb. Janus yellow R and 8 oz. Janus grey B B in the dye-bath, with 12 lb. sumac extract and 1¼ lb. tartar emetic in the fixing-bath.

Chestnut Brown. — 1 lb. Janus brown R and 1 lb. Janus yellow R in the dye-bath, and 8 lb. sumac extract and 1 lb. tartar emetic in the fixing-bath.

Before the introduction of the direct dyes the method usually followed, and, indeed still used to a great extent, is that known as cross dyeing. The goods were woven with dyed cotton threads of the required shade, and undyed woollen threads. After weaving and cleansing the woollen part of the fabric was dyed with acid dyes, such as Acid magenta, Scarlet R, Acid yellow, etc. In such methods care has to be taken that the dyes used for dyeing the cotton are such as stand acids, a by no means easy condition to fulfil at one time. Many of the direct dyes are fast to acids and, therefore, lend themselves more or less readily to cross dyeing. For details of the dyes for cotton reference may be made to the sections on dyeing with the direct colours, page 85, etc., while information as to methods of dyeing the wool will be found in the companion volume to this on *Dyeing of Woollen Fabrics.*

Shot Effects. — A pleasing kind of textile fabric which is now made, and is a great favourite for ladies' dress goods, is where the cotton of a mixed fabric is thrown up to form a figured design. It is possible to dye the two fibres in different colours, and so produce a variety of shot effects. These latter are so endless that it is impossible here to enumerate all that may be produced. It will have to suffice to lay down the lines which may be followed to the best advantage, and then give some recipes to illustrate the remarks that have been made. The best plan for the production of shot effects upon union fabrics is to take advantage of the property of certain acid dyes which dye only the wool in [Pg 221] an acid bath, and of many of the direct colours which will only dye the cotton in an alkaline bath. The process, working on these lines, becomes as follows: The wool is first dyed in an acid bath with the addition of Glauber's salt and bisulphate of soda, or sulphuric acid, the goods are then washed with water containing a little ammonia to free them from the acid, and afterwards dyed with the direct colour in an alkaline bath.

Fancy or the mode shades are obtained by combining suitable dye-stuffs.

If the cotton is to be dyed in light shades it is advantageous to dye on the liquor at 65° to 80° F., with the addition of 3¼ oz. Glauber's salt, and from 20 to 40 grains borax per gallon water. The addition of an alkali is advisable in order to neutralise any slight quantities of acid which may have remained in the wool, and to prevent the dye-stuff from dyeing the cotton too deep a shade.

Very light shades can also be done on the padding machine. The dye-stuffs of Group II., which have been previously enumerated, do not stain the wool at all, or only very slightly, and are, therefore, the most suitable. Less bright effects can be produced by simply dyeing the goods in one bath. The wool is first dyed at the boil with the respective wool dye-stuff in a neutral bath, the steam is then shut off and the cotton dyed by adding the cotton dye-stuff to the bath, and dyeing without again heating. By passing the goods through cold water to which some sulphuric or acetic acid is added, the brightness of most effects is greatly increased.

Gold and Green.—First bath, 1 lb. Cyanole extra, 7¼ oz. Acid green, 1½ oz. Orange G G, and 10 lb. bisulphate of soda; work at the boil for one hour, then lift and rinse well. Second bath, 4 lb. Diamine orange G and 15 lb. Glauber's salt; work in the cold or at a luke-warm heat. Third bath, at 120° F., 4 oz. Chrysoidine and ¼ oz. Saf-ranine. [Pg 222]

Black and Blue.—First bath, 3½ lb. Naphthol black 3 B and 10 lb. bisulphate of soda. Second bath, 2 lb. Diamine sky blue and 13 lb. Glauber's salt. Third bath, 6½ oz. New Methylene blue N. Work as in the last recipe.

Green and Claret.—First bath, 3½ lb. Naphthol red C and 10 lb. bi-sulphate of soda. Second bath, 2 lb. Diamine sky blue F F, 1¼ lb. Thioflavine S, and 15 lb. Glauber's salt.

Gold Brown and Blue. — First bath, 2½ oz. orange E N Z, 1½ oz. Orange G G, ¼ oz. Cyanole extra and 10 lb. bisulphate of soda. Second bath, 14 oz. Diamine sky blue F F and 15 lb. Glauber's salt.

Dark Brown and Blue. — First bath, ½ lb. Orange G G, 1½ oz. Orange E N Z, 1½ oz. Cyanole extra and 10 lb. bisulphate of soda. Second bath, 12 oz. Diamine sky blue F F and 15 lb. Glauber's salt.

Black and Green Blue. — First bath, 3 lb. Orange G G, 1 lb. Brilliant cochineal 4 R, 1 lb. Fast acid green B N and 10 lb. Glauber's salt. Second bath, 1¾ lb. Diamine sky blue F F, 3¼ lb. Thioflavine S and 15 lb. Glauber's salt.

We may here note that in all the above recipes the second bath (for dying the cotton) should be used cold or at lukewarm heat and as strong as possible. It is not completely exhausted of colour, only about one-half going on the fibre. If kept as a standing bath this feature should be borne in mind, and less dye-stuff used in the dying of the second and following lots of goods.

Blue and Gold Yellow. — 3 lb. Diamine orange G, 13 oz. Naphthol blue G, 14½ oz. Formyl violet S 4 B and 15 lb. Glauber's salt. Work at just under the boil.

Brown and Blue. — 1 lb. Diamine steel blue L, 9½ oz. Diamine sky blue, 1 lb. Orange E N Z, 1 lb. Indian yellow G, 1¾ oz. Naphthol blue black and 15 lb. Glauber s salt. Work at 170° to 180° F.

In these two last recipes only one bath is used, all the [Pg 223] dyes being added at once. This is possible if care be taken that dye-stuffs of two kinds are used, one or more which will dye wool and not cotton from neutral baths, and those direct dyes which dye cotton better than wool. The temperature should also be kept below the boil and carefully regulated as the operation proceeds and the results begin to show themselves.

Grey and Orange. — First bath, 3 oz. Orange extra, 1¼ lb. Cyanole extra, 1 lb. Azo red A and 10 lb. bisulphate of soda. Second bath, 5 oz. Diamine orange D C and 3 oz. Diamine fast yellow B.

Green and Red. — First bath, 2 lb. Croceine A Z, and 10 lb. Glauber's salt. Second bath, 1 lb. Diamine sky blue F F, ½ lb. Thioflavine S, and 15 lb. Glauber's salt.

Brown and Violet.—First bath, ¾ lb. Orange extra, ¾ lb. Cyanole extra, and 10 lb. bisulphate of soda. Second bath, 5 oz. Diamine brilliant blue G, and 15 lb. Glauber's salt.

Black and Yellow.—First bath, 7 lb. Naphthol black B, ½ lb. Fast yellow S, and 10 lb. bisulphate of soda. Second bath 3 lb. Diamine fast yellow A, and 15 lb. Glauber's salt.

Black and Pink.—Black as above. Pink with Diamine rose B D (see above).

Green and Buff.—First bath, ¼ lb. Orange extra, ¾ oz. Fast yellow S, and 10 lb. bisulphate of soda. Second bath, ¾ lb. Diamine sky blue F F, ½ lb. Thioflavine S, and 15 lb. Glauber's salt.

Orange and Violet.—First bath, 9 oz. Orange extra, and 10 lb. bisulphate of soda. Second bath, ¾ lb. Diamine violet N, and 10 lb. Glauber's salt.

Black and Blue.—First bath, Naphthol black as given above. Second bath, Diamine sky blue as given above.

Black and Yellow.—Add first 1 lb. Wool black 6 B, and 10 lb. Glauber's salt, then, when the wool has been dyed, add 2 lb. Curcumine S to dye the cotton in the same bath. [Pg 224]

Green and Red.—Dye the wool by using 3 lb. Guinea green B, ¼ lb. Curcumeine extra, and 10 lb. Glauber's salt, then add to the bath ¾ lb. Erika B N, and ¾ lb. Congo Corinth G.

Orange and Blue.—Dye the wool first with 1¼ lb. Mandarine G, 2 oz. Wool black 6 B, and 10 lb. Glauber's salt; then the cotton with 2 lb. Columbia blue G.

Blue and Orange.—Dye the wool first with ¾ lb. Guinea violet B, ¾ lb. Guinea green B, and 10 lb. Glauber's salt; then dye the cotton with 2 lb. Mikado orange 4 R O.

Green and Orange.—Dye the wool with 3 lb. Guinea green B, ¼ lb. Curcumeine extra, and 10 lb. Glauber's salt, then dye the cotton in the same bath with 1½ lb. Mikado orange 4 R O. [Pg 225]

CHAPTER VI.

DYEING HALF SILK (COTTON-SILK, SATIN) FABRICS.

The direct dyes of the Diamine, Benzo and Congo types have been of late years increasingly used for dyeing satin (silk and cotton), and they have quite displaced the old methods of dyeing this class of fabrics, which consisted in first dyeing the silk with an acid dye and then dyeing the cotton with a basic dye. For details of the method of applying acid dyes to silk reference may be made to Mr. G.H. Hurst's book on *Silk Dyeing*.

Most of the direct colours are exceedingly well adapted for this purpose, some under certain conditions possess the property of dyeing the cotton a deeper shade than the silk, which is an advantage rather than otherwise.

The dyeing of goods composed of silk and cotton is generally done in winch dye-vats, in some cases also on the jigger.

METHOD OF DYEING.

The direct colours are as a rule dyed in a soap-bath with addition of phosphate of soda, Glauber's salt or common salt and a little soda.

The addition of these salts effects a better exhaustion of the baths; they are therefore principally used for dark and full shades, whilst pale shades are dyed with the addition of soap only or in combination with phosphate of soda. Dark or pale shades may thus be produced at will by selecting the [Pg 226] proper additions, but the fact should not be overlooked that the greater exhaustion of the baths not only increases the depth of shade of the cotton but also causes the silk to absorb more dye-stuff. Too large a proportion of salt would cause the dye-stuffs to go on the fibre too quickly and thus make the dyeing liable to turn out uneven.

A large proportion of soap counteracts the effects of the salts, causing the dye-stuff to go on less quickly and tending to leave the silk lighter than the cotton, in some cases even almost white, a property which is valuable in many cases, especially as enabling the silk and cotton to be dyed in different colours to obtain shot effects.

It is thus obvious that a general method applicable in all cases cannot be given; it will vary according to the effect desired, and partly also depend on the material to be dyed.

The following particulars may serve as a guide for the first bath:—

For pale shades each 10 gallons dye-liquor should contain 3¼ to 6½ oz. soap and 4 to 7 drs. soda or 3¼ to 6½ oz. soap, 4 to 5½ drs. soda and 3¼ to 6½ oz. phosphate of soda.

For medium and dark shades each 10 gallons dye-liquor may contain 3¼ to 6½ oz. soap, 4 to 7 drs. soda, 3¼ to 6½ oz. phosphate of soda and 6½ to 13 oz. cryst. Glauber's salt.

For two coloured effects or dyeings, in which the silk is intended to remain as pale as possible or even white, each 10 gallons dye-liquor may contain 4¾ to 8 oz. soap, 4 to 6 drs. soda, 3¼ to 8 oz. phosphate of soda and 4¾ to 9½ oz. cryst. Glauber's salt.

The temperature of the dye-baths is generally 175° to 195° F.; in practical dyeing it is usual to boil up the fully charged dye-bath, shut off the steam, enter the goods and dye for about three-quarters of an hour.

For obtaining level dyeings in pale shades it is advisable not to enter the goods too hot, but to raise the temperature [Pg 227] gradually. Raising the temperature, or dyeing for some time at the boil will deepen the shade of the cotton, but at the same time will have the same effect on the silk which may sometimes be an advantage when dyeing dark shades.

As a complete exhaustion of the baths does not take place, especially when dyeing dark shades, it is advantageous, nay, even imperative, to preserve the baths for further use, they are then replenished with only about three-fourths of the quantities of dye-stuffs used for the first bath, of the soap only about one fourth, of Glauber's salt, soda and phosphate of soda only about one-fifth, of the first quantities are necessary.

The first bath should be prepared with condensed water. If none is at hand ordinary water should be boiled up with soda and soap and the scum removed. Clear soap baths are absolutely necessary for the production of pure shades and clean pieces.

After dyeing, the pieces must be very well rinsed, and the colour raised or brightened with 1 pint of acetic acid in 10 gallons of water.

Many of the Diamine and Titan colours being very fast to acids, but few of them will be affected by this treatment.

In the following tables are given those Diamine, etc., colours especially adapted for the dyeing of goods composed of silk and cotton, divided into three groups according to their relation to silk and cotton: —

1. Dye-stuffs possessing a great affinity to cotton and tinting the silk not at all or only very little. To this class belong Chicago blues, Benzo blues, Diamine fast yellow A, Diamine orange G G, Diamine orange D C, Diamine blue B B, Diamine blue 3 B, Diamine sky blue F F, Diamine brilliant blue G, Diamineral blue E, Diamine black B, Mikado browns, Mikado oranges, Mikado yellows.

[Pg 228]

2. Dye-stuffs producing on cotton and silk the same or nearly the same shade but covering the cotton better than the silk. These are Thioflavine S, Diamine yellow N, Diamine gold, Diamine fast yellow B, Diamine orange B, Diamine grey G, Diamine rose B D, Diamine scarlet S, Diamine scarlet B, Diamine scarlet 3 B, Diamine red 5 B, Diamine fast red F, Diamine Bordeaux B, Diamine Bordeaux S, Diamine violet N, Oxydiamine violet B, Diamine blue R W, Diamine black H W, Diamine steel blue L, Diamine dark blue B, Union black S, Oxydiamine black D, Diaminogene extra, Diaminogene B, Diamine brown M, Diamine brown 3 G, Diamine green B, Diamine green G.

3. Dye-stuffs producing on cotton more or less different shades than on silk. This group comprises Diamine blue C B, Diamine blue B G, Diamine blue B X, Diamine azo blue 2 R, Diamine blue 3 R, Diamine blue black E, Diamine black R O, Oxydiamine black S O O O, Diamine brown V, Diamine brown B, Diamine bronze G. Cotton brown N produces on silk darker shades than on cotton.

Of course this classification cannot be taken as absolutely correct, as by raising or lowering the temperature during the dyeing process or by a larger or smaller addition of soap or Glauber's salt (common salt, phosphate of soda), the dye-stuffs are more or less influenced in one or the other direction. Diamine violet N, for instance, when dyed with an increased addition of soap would dye the cotton somewhat lighter, but at the same time leave the silk perfectly white.

Topping with Basic and Acid Dye-stuffs. — As in very few cases only the desired shade can be obtained in the first instance by bottoming with direct colours, topping generally has to be resorted to. This is best done with basic dyes, in some cases also with acid dye-stuffs in cold or tepid bath with addition of sulphuric acid, hydrochloric or acetic acid. The use of acid dye-stuffs is restricted to cases where the silk [Pg 229] alone is to be shaded. In most cases basic dye-stuffs are made use of, which dye silk and cotton the same shade and deepen the shade of the cotton if the latter has a sufficiently good bottom, thus giving the goods a better and fuller appearance.

It is not advisable to employ basic and acid dye-stuffs in the same bath except when the quantities of either class are very small. Should it be necessary to dye with large quantities of both classes, the acid dye-stuffs are first dyed in a tepid acid bath and then the goods are topped with the basic dye-stuffs in a fresh cold bath with the addition of a little hydrochloric or acetic acid.

Of the basic dye-stuffs which are available, the following are the most suitable for topping: New methylene blue N, and other brands; New blue D and other brands; Cresyl blue, Methylindone B and R, Metaphenylene blue, Indazine; the various brands of Brilliant green, Solid green and Malachite green, Capri green, Cresyl violet, Thioflavine T, New phosphine G, Tannin orange R, and the various brands of Bismarck brown; Safranine, Magenta all brands, Tannin heliotrope, all brands of Neutral violet, Methyl violet.

Of the acid dye-stuffs, the following are good for topping or shading the silk: Cyanole extra, Indigo blue N, Indigo blue S G N, and the various brands of Water blue, Soluble blue, Solid blue, and Induline; the various brands of Acid green and Fast acid green; Indian yellow G and R, Naphthol yellow S, Tropæoline O and O O, and the various brands of Milling yellow and Orange; Azo red A, Azo rubine A, Archil substitute N, Azo orseille B B, Brilliant orseille C, and the various brands of Eosine, Erythrosine, Rose bengale, Rhodamine, Brilliant croceine and Brilliant scarlet; the various brands of

Formyl violet and Acid violet; Aniline grey B and Nigrosine, soluble in water.

Bright Yellow.—Use 2 lb. Thioflavine S. [Pg 230]

Deep Orange Yellow.—This can be dyed by using 2 lb. Diamine yellow N.

Gold Yellow.—Dye with 2 lb. Diamine gold. Some care must be taken with this, especially not to dye too hot or the silk will be dyed deeper than the cotton.

Deep Orange.—Use 2 lb. Diamine orange B.

Bright Rose.—Use 2 lb. Diamine Rose B D. Do not work too high, especially when dyeing light rose shades, as then the silk is apt to take up too much colour.

Scarlet.—Use in the dye-bath 2 lb. Diamine scarlet H S. The heat of the dye-bath should not be allowed to exceed 160° to 170° F., or there is a risk of the shades becoming somewhat duller.

Crimson.—Dye with 2 lb. Diamine fast red F.

Violet.—Use 2 lb. Oxydiamine violet B.

Bright Blue.—A fine shade is dyed with 2 lb. Diamine blue R W.

Dark Green.—Use 2 lb. Diamine black H W. This gives a fine shade of bluish green.

Gold Brown.—Dye with 2 lb. Diamine brown 3 G at a low heat, from 150° to 160° F., otherwise the silk takes up too much colour.

Dark Green.—Dye with 2 lb. Diamine green B.

Deep Rose.—Dye with 2 lb. Diamine red 10 B.

Brilliant Yellow.—Dye with 1½ lb. Mikado golden yellow 8 G; then enter into a cold bath which contains 1½ per cent. Auramine II. This gives a very bright shade of yellow.

Dark Brown.—Dye a bottom with 2 lb. Mikado brown 3 G O, and then top with 3 lb. Bismarck brown and ½ lb. Capri blue G O N.

Crimson.—Dye with 2 lb. Mikado orange 5 R O and 2 lb. Hessian purple N.

Sage Green. — Dye a bottom with 2 lb. Mikado yellow G, 14 oz. Eboli green T and 3 oz. Mikado brown M, then top in [Pg 231] a fresh cold bath with ½ lb. Auramine II and ½ oz. Acridine Orange N 0.

Leaf Green. — Dye a bottom with 3 lb. Mikado golden yellow 8 G and 1 lb. Eboli blue B; then top with 1½ lb. Capri green 2 G in a cold bath.

Deep Brown. — Dye with 2 lb. Mikado orange 3 R O, 3 lb. Hessian grey S and 1 lb. Hessian brown 2 B N; then top with 7 oz. Azine green T 0 and 2¼ lb. Acridine orange N 0.

Dark Cream. — Bottom with 1 oz. Diamine orange G; then top in a fresh warm bath with 1 oz. Orange G G, ½ oz. Indian yellow R, 5 lb. Glauber's salt and 1 lb. acetic acid.

Brilliant Violet. — Give a bottom with 1 lb. Diamine violet N; then top in a fresh warm bath with 4 oz. Methyl violet B and 2 oz. Rhodamine.

Slate. — Bottom in a hot bath with 6 oz. Diamine dark blue B and 1½ oz. Diamine brown M; then top in a fresh bath at 170° F. with 4 oz. Aniline grey B, 1 oz. Cyanole extra, 5 lb. Glauber's salt and 1 lb. acetic acid.

Black Brown. — Give a bottom with 2 lb. Cotton brown A, 1 lb. Diamine gold and 3½ lb. Oxydiamine black S O O O; then top in a fresh bath at 120° F. with 4 oz. New methylene blue N, 1 oz. Safranine and ½ oz. Indian yellow G.

Bright Violet. — Use ½ lb. Oxydiamine violet B and ¾ oz. Diamine dark blue B; top after dyeing with ½ oz. Safranine, ¼ oz. Methylindone B and ¼ oz. Cyanole extra.

Drab. — Dye with 6 oz. Diamine orange G, 1 lb. Diamine bronze G and ¾ lb. Diamine brown M, topping afterwards in a bath of ¼ oz. Aniline grey B and ¼ oz. Bismarck brown F F.

Leaf Green. — Dye with ½ lb. Diamine black H W, and 1 lb. Diamine fast yellow B; top with ¼ oz. Brilliant green, ¼ oz. Indian yellow R, ½ oz. Thioflavine T and ½ oz. Cyanole extra.

Dark Crimson.—Use in the dye-bath 3 lb. Diamine Bordeaux S, ¾ lb. Diamine orange D C and 1½ lb. Diamine brown V, topping with 1 oz. Magenta and ½ oz. Formyl violet S 4 B. [Pg 232]

Turquoise Blue.—Use to dye the ground, 6 oz. Diamine sky blue F F and ½ oz. Diamine fast yellow A; top with 1½ oz. Cyanole extra and ¼ oz. Brilliant green.

Dark Grey.—Dye with ½ oz. Diamine grey G, and 1½ oz. Diamine brown M; top with ¼ oz. Orange extra and 1 oz. Cyanole extra.

Brilliant Orange.—Dye with 1 lb. Mikado orange R O, and top with 6 oz. Acridine orange N O and 12 oz. Auramine I I.

Brown.—Dye a bottom colour with 3 lb. Mikado brown M, and top with 2 lb. Bismarck brown and 6 oz. Cresyl fast violet 2 R N.

Deep Crimson.—Dye with 1 lb. Columbia black R and top with 6 oz. Magenta.

Pale Sea Green.—Use in the dye-bath ½ oz. Chrysophenine G, 1½ oz. Chicago blue 6 B and 1½ oz. Alkali blue 6 B.

Bright Crimson.—Dye with 3 lb. Congo Corinth and top with 1 lb. Magenta.

Dark Russian Green.—Dye with 3 lb. Columbia black B; then top with 1 lb. Malachite green.

Gold Drab.—Dye with 5 oz. Columbia black, and top with 5 oz. Chrysoidine R.

Bright Olive Yellow.—Dye with 1½ lb. Diamine gold, 1½ lb. Diamine fast yellow A and ¾ lb. Diamine bronze G; top with ½ lb. Thioflavine T and ¼ lb. Chrysoidine.

Moss Brown.—Dye with 1 oz. Diamine brown M, 6 oz. Diamine fast yellow A, 6 oz. Diamine bronze G, topping with 1 oz. new Methylene blue N and 4 oz. Orange G G.

Dark Sea Green—Dye a bottom with 9 oz. Diamine black B and 4½ oz. Diamine fast yellow B, then top with 2 oz. New methylene blue M and 2 oz. New phosphine G.

Old Gold.—Dye a ground with ½ lb. Diamine gold, 1¼ lb. Diamine fast yellow A, and 6 oz. Diamine bronze G, topping with 8 oz.

Thioflavine T, 1 oz. Indian yellow R and 1 oz. Brilliant green. [Pg 233]

Cornflower Blue. — Dye the ground with 2½ lb. Diamine azo blue 2 B, 1½ oz. Alkali blue 3 B, ½ lb. Oxydiamine black S O O O, and top with 1 oz. Metaphenylene blue B, 2 oz. New methylene blue R and 1 oz. Indigo blue N.

Slate. — Dye with 7 oz. Diamine dark blue B and 1 oz. Diamine brown M; top with 1 oz. Aniline grey B and 1 oz. Cyanole extra.

Pale Drab. — Dye the ground with 1 oz. Diamine orange G C, ¾ oz. Diamine bronze G and ½ oz. Diamine brown M; top with ¾ oz. New methylene blue N, 1 oz. Bismarck brown and 1 oz. Cyanole extra.

Deep Leaf Green. — Dye a ground colour with 1¼ lb. Diamine bronze G, 1½ lb. Diamine fast yellow A and 1½ lb. Diamine black H W; the topping bath is made with ½ lb. Brilliant green, ½ lb. Chrysoidine and ¼ lb. New methylene blue N.

Maroon. — Dye with 3 lb. Diamine Bordeaux S, ½ lb. Diamine orange D C and ½ lb. Diamine brown V; top with ½ lb. Magenta and ¼ lb. Formyl violet S 4 B.

Heliotrope. — Dye with 1 lb. Heliotrope 2 B.

Lilac Rose. — Dye with 8 lb. Columbia black R and 1 lb. Alkali blue B; after dyeing pass through a weak acetic acid bath, then wash well.

Pea Green. — Dye with 2 lb. Chrysophenine, 1 lb. Chicago blue 6 B and 1 lb. Alkali blue 6 B; pass, after dyeing, through a weak acetic acid bath, then wash well.

Dark Drab. — Dye with ¼ lb. Diamine brown M, 1 lb. Diamine fast yellow A and ¾ lb. Diamine bronze G; top with ½ lb. Orange G G and ½ lb. Cyanole extra.

Deep Rose. — Dye the bottom colour with ½ lb. Diamine rose B D and top with ¼ lb. Rhodamine B and 1 oz. Safranine.

Walnut Brown. — Dye the bottom colour with 1 lb. Oxydiamine black D, 1 lb. Diamine brown M and 1 lb. Oxydiamine violet B; the topping is done with 4 oz. Safranine, 2 oz. New methylene blue N and 2 oz. Chrysoidine. [Pg 234]

Dyeing of Plain Black.—Diamine blacks find a very extensive application for dyeing blacks on satin, either dyed direct in one bath, or dyed, diazotised and developed.

Union black S and Oxydiamine black D are particularly suitable for direct blacks, and are used either alone or in a combination with Diamine jet black S S, which produces a better covering of the silk, or with Oxydiamine black S O O O, which deepens the shade of the cotton. According to the shade required Diamine fast yellow A and B, Diamine green B or G, or Alkaline blue may be used for shading.

Dye for about one hour at about 175° to 195° F. in as concentrated a bath as possible, with about 7 to 8 lb. dye per 100 lb. of satin, 8 to 16 oz. Glauber's salt and 5 to 8 oz. soap per 10 gallons dye liquor; keep cool in the bath for some time and rinse.

The raising is either done in a tepid soap bath with the addition of some new methylene blue, or in an acid bath to which Naphthol, blue black, Acid green, etc., is added for shading the silk.

Direct dyed blacks are especially suitable for cheap goods (ribbons, light linings, etc.), for which special fastness to water is not required; also for tram and tussar silk plushes, which are afterwards topped with logwood.

If greater fastness is required, and more especially if it is a case of replacing aniline black, Diaminogene diazotised and developed is a good dye-stuff. It is extensively used for dyeing umbrella cloths and linings. Against aniline black it has the great advantage of not tendering the fibre in the least, and not turning green during storage. Diaminogene B and Diaminogene extra are mostly used for this purpose, the former for jet blacks, the latter for blue-black shades.

Proceed as follows: Enter the boiled off and acidulated goods in a boiling bath as concentrated as possible, charged [Pg 235] with 16 oz. Glauber's salt per 10 gallons liquor, and 1 lb. acetic acid per 100 lb. dry goods. For jet black add for 100 lb. satin, 6 to 8 lb. Diaminogene, 1 to 2 lb. Naphthylamine black D, ½ to 1 lb. Diamine fast yellow A or Diamine green B; for very deep shades about 1/5 of the quantity of Diaminogene B may be replaced with Diamine jet black

S S. For blue black, 6 to 8 lb. Diaminogene B, or 3 to 4 lb. Diamino-gene B, and 3 to 4 lb. Diaminogene extra. Dye for three-quarters to one hour at the boil, allow to cool in the bath for about thirty minu-tes, then rinse, diazotise and develop.

Phenylene diamine (93 per cent.) serves for developing jet blacks mixed with resorcine for greenish shades. Beta-naphthol is used for blue blacks (1 lb. 5 oz. per 100 lb. of dry material, dissolved in its own weight of soda lye, 75° Tw.). The three developers may also be mixed with each other in any proportions.

After developing soap hot with addition of new methylene blue, by choosing a reddish or a bluish brand of new methylene, blue and black may be shaded at will in the soap bath; finally rinse and raise with acetic acid.

If properly carried out this process will give a black almost equal to aniline black; but having, as already mentioned, the advantage of not impairing the strength of the fibre, and not turning green during storage.

As the dye-baths for blacks are charged with a proportionately high percentage of dye-stuff for the first bath, and will not exhaust completely, it is advisable to preserve them for further use.

For subsequent lots only two-thirds to three-fourths of the quanti-ties of dye-stuffs used for the first baths are required, which fact has to be taken into consideration when calculating the cost of dyeing.

Dyeing Shot Effects on Satin. — Not all direct colours are equally well adapted for the production of shot effects; [Pg 236] those enu-merated in Group I. are most suitable for the purpose, and should be dyed with a larger quantity of soap than is usual for solid shades, in order to leave the silk as little tinted as possible. Dye-stuffs of the other groups may be used if the dyeing is conducted with proper care, *i.e.*, keeping the baths more alkaline and lowering the tempera-ture. The goods are dyed with the addition for the two coloured effects previously mentioned, then they are well rinsed, and after-wards the silk is dyed with the suitable acid dye-stuffs, with additi-on of sulphuric acid at a temperature of about 150° F. Care should be taken not to use too much acid, and to keep the temperature of the bath sufficiently low, as otherwise the acid may cause some of the dye-stuff to go off the cotton and tint the silk. It is best to work at a temperature of about 150° F., with addition of about 3 oz. concentrated sulphuric acid per 10 gallons dye-liquor.

For shading the silk all acid dye-stuffs can be used which have been mentioned in the foregoing tables.

If in shot effects the cotton is to be dyed bright and full shades, this is best achieved by dyeing with direct colours first, and then topping with basic colours as follows: —

Bottom the cotton first with the suitable direct colours, then dye the silk and then treat the pieces for about two hours in a cold tannin bath (about 8 oz. tannin per 10 gallons of water), then rinse once and pass through a tartar emetic bath (about 3 oz. per 10 gallons), rinse thoroughly and dye the cotton to shade with basic colours in a cold bath to which some acetic acid has been added.

Should the silk become a little dull after this process, this may be remedied by a slight soaping. After dyeing rinse well and raise with acetic acid.

Shot Effects with Black Cotton Warp.—Effects much in favour are designs composed of black cotton and light or [Pg 237] coloured silk. The most suitable black dye for this purpose is Diamine black B H, diazotised and developed.

Dye in as concentrated a bath as possible at about 160° F. with about 6 lb. Diamine black B H, 1 lb. Diamine sky blue, pat., per 100 lb. of dry goods, ½ lb. Diamine orange D C, pat., with an addition of 6½ oz. soap, 4 to 5 dr. soda per 10 gallons liquor, 16 oz. Glauber's salt. After dyeing rinse well in a bath containing 6 dr. soda and 3 oz. soap per 10 gallons water, diazotised in a fresh bath with 4 lb. nitrite of soda and 12 lb. hydrochloric acid (per 100 lb. of dry goods), rinse thoroughly and develop with 3 to 16 oz. phenylene diamine (93 per cent.), with addition of 1 to 2 lb. soda. These two operations should follow each other as quickly as possible, also care has to be taken that the diazotised goods are not exposed to direct sunlight or heat, which causes unlevel dyeings. The silk is then cleaned as far as possible by hot soaping, and dyed at about 120° to 140° F., with acid dye-stuffs and the addition of sulphuric acid. After dyeing rinse as usual and brighten.

Yellow and Violet,—Dye the cotton with 2 lb. Diamine fast yellow A, the silk with 1 lb. Cyanole extra, and 1 lb. Forinyl violet S 4 B.

Black and Blue.—Dye the cotton with 5 lb. Diamine black B H, 1 lb. Diamine sky blue, and ¼ lb. Diamine orange D C. After dyeing, diazotise and develop with phenylene diamine as described above. Then dye the silk with ½ lb. Pure soluble blue and 1 lb. Cyanole extra.

Black and Crimson.—Dye the black as in the previous recipe, then dye the silk with 2 lb. Brilliant croceine 3 B and ½ lb. Rhodamine S.

Blue and Gold.—Dye the cotton with 2 lb. Diamine sky blue and the silk with 1 lb. Fast yellow S.

Dark Blue and Green.—Dye the cotton with 1½ lb. Diamine black B H, 1½ lb. Diamine sky blue and ½ lb. Diamine azo [Pg 238] blue 2 R; the silk with 2 lb. Naphthol yellow S and 1 lb. New methylene blue G G.

Violet and Yellow.—Dye the cotton with 2 oz. Diamine violet N and the silk with 1 lb. Fast yellow S.

Orange and Violet.—Dye the cotton with 2 lb. Diamine orange D C and the silk with 1 lb. Formyl violet S 4 B.

Dark Blue and Olive.—Dye the cotton with 1½ lb. Diamineral blue R and ½ lb. Diamine azo blue 2 R, and the silk with 1 lb. Naphthol yellow B and 1 lb. Orange G G.

Green and Pink.—Dye the cotton with 1½ lb. Diamine fast yellow A and ¼ lb. Diamine sky blue, and the silk with 1 lb. Erythrosine B.

Brown and Blue.—Dye the cotton with 3 lb. Mikado brown 2 B, and the silk with ½ lb. Pure blue.

It is quite possible to produce two coloured effects containing blue in one bath by using Alkali blue as a constituent with a direct dye which works only on to the cotton, the alkali blue going on to the silk, as, for example, in the following recipes:—

Orange and Blue.—The dye-bath is made with 3 lb. Mikado orange 5 R O and 1¼ lb. Alkali blue 6 B. After the dyeing the goods are rinsed, then passed through a bath of 1½ lb. sulphuric acid in 10 gallons water, washed well and dried.

Olive and Blue.—The dye-bath is made with 1½ lb. Diamine fast yellow A, 2½ lb. Diamine orange DC, ¼ lb. Diamine sky blue, and 1 lb. Alkali blue 6 B, After dyeing rinse, then acidulate as above and wash well. [Pg 239]

CHAPTER VII.

OPERATIONS FOLLOWING DYEING.

WASHING, SOAPING, DRYING.

After loose cotton or wool, or cotton and woollen yarns, or piece goods of every description have been dyed, before they can be sent out for sale they have to pass through various operations of a purifying character. There are some operations through which cloths pass that have as their object the imparting of a certain appearance and texture to them, generally known as finishing processes; of these it is not intended here to speak, but only of those which precede these, but follow on the dyeing operations.

These processes are usually of a very simple character, and common to most colours which are dyed, and here will be noticed the appliances and manipulation necessary in the carrying out of these operations.

Squeezing or Wringing.—It is advisable when the goods are taken out of the dye-bath to squeeze or wring them according to circumstances, in order to press out all surplus dye-liquor, which can be returned to the dye-bath if needful to be used again. This is an economical proceeding in many cases, especially in working with many of the old tannin materials like sumac, divi-divi, myrobolams, and the modern direct dyes which in the dyeing operations are not completely extracted out of the bath, or in other words, the dye-bath is not exhausted of colouring matter, and, therefore, it can [Pg 240] be used again for another lot of goods, simply by adding fresh material to make up for that absorbed by the first lot of goods.

Loose wool and loose cotton are somewhat difficult to deal with by squeezing or wringing, but the material may be passed through a pair of squeezing rollers, such as are shown in Fig. 31, which will be more fully dealt with later on. The machine shown is made by Messrs Read Holliday & Sons.

FIG. 31.—Squeezing Rollers.

Yarns in Hanks.—In the hank-dyeing process the hanks are wrung by placing one end of the hank on a wringing [Pg 241] horse placed over the dye-tub, a dye stick on the other end of the hank giving two or three sharp pulls to straighten out the yarn, and then twisting the stick round, the twisting of the yarns puts some pressure on the fibres, thoroughly and uniformly squeezing out the surplus liquor from the yarn.

Hank-wringing Machines.—Several forms of hank-wringing machines have been devised. One machine consists of a pair of discs fitted on an axle; these discs carry strong hooks on which the hanks are placed. The operator places a hank on a pair of the hooks. The

224

discs revolve and carry round the hank, during the revolution the hank is twisted and the surplus liquor wrung out, when the revolution of the discs carries the hank to the spot where it entered the machine the hooks fly back to their original position, the hank unwinds, it is then removed and a new hank put in its place, and so the machine works on, hanks being put on and off as required. The capacity of such a machine is great, and the efficiency of its working good.

Mr. S. Spencer of Whitefield makes a hank-wringing machine which consists of a pair of hooks placed over a vat. One of the hooks is fixed, the other is made to rotate. A hank hung between the hooks is naturally twisted, and all the surplus liquor wrung out. The liquor falling into the vat.

Roller Squeezing Machines for Yarn.—Hanks may be passed through a pair of indiarubber squeezing rollers, which may be so arranged that they can be fixed as required on the dye-bath. Such a pair of rollers is a familiar article, and quite common and in general use in dye houses.

Piece Goods.—These are generally passed open through a pair of squeezing rollers which are often attached to the dye-vat in which the pieces are dyed.

Read Holliday's Squeezing Machine.—In Fig. 31 is shown a squeezing machine very largely employed for squeezing [Pg 242] all kinds of piece goods and cotton warps after dyeing or washing. It consists of a pair of heavy rollers on which, by means of the screws shown at the top, a very considerable pressure can be brought to bear. The piece is run through the eye shown on the left, by which it is made into a rope form, then over the guiding rollers and between the squeezing rollers, and into waggons for conveyance to other machines. This machine is effective.

Another plan on which roller, or rather in this case disc, squeezing machines is made, is to make the bottom roller with a square groove in the centre, into this fits a disc, the cloth passing between them. The top disc can by suitable screws be made to press upon the cloth in the groove, and thus squeeze the water out of it.

Washing.—One of the most important operations following that of dyeing is the washing with water to free the goods, whether cotton or woollen, from all traces of loose dye, acids, mordanting materials, etc., which it is not desirable should be left in, as they might interfere with the subsequent finishing operations. For this purpose a plentiful supply of good clean water is required; this should be as soft as possible, free from any suspended matter which might settle upon the dyed goods, and stain or speck them.

Washing may be done by hand, as it frequently was in olden days, by simply immersing the dyed fabrics in a tub of water, shaking, then wringing out, again placing in fresh water to finish off. Or if the dye-works were on the banks of a running stream of clean water the dyed goods were simply hung in the stream to be washed in a very effectual manner.

In these days it is best to resort to washing machines adapted to deal with the various kinds of fibrous materials and fabrics in which they can be subjected to a current of water. [Pg 243]

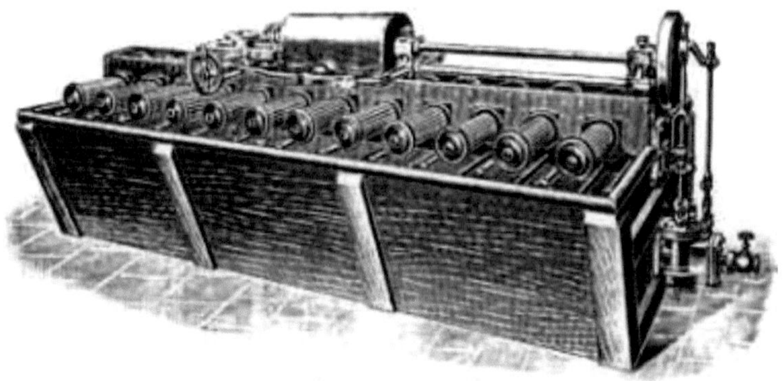

FIG. 32.—Yarn-washing Machine.

[Pg 244]

Loose Wool and Cotton.—If this has been dyed by hand then the washing may also be done in the same way by hand in a plain vat. If the dyeing has been done on a machine, then the washing can be done on the same machine.

Yarns.—Yarn in the cop form is best washed in the machine in which it is dyed.

Yarns in Hanks.—A very common form of washing machine is shown in Fig. 32. As will be seen it consists of a wooden vat, over which are arranged a series of revolving reels on which the hanks are hung. The hanks are kept in motion through the water, and so every part of the yarn is thoroughly washed. Guides keep the hanks of yarn separate and prevent any entanglement one with another. A pipe delivers constantly a current of clean water, while another pipe carries away the used water. Motion is given to the reels in this case by a donkey engine attached to the machine, but it may also be driven by a belt from the main driving shaft of the works. This machine is very effective.

FIG. 33.—Dye-house Washing Machine.

Piece Goods.—Piece goods are mostly washed in machines, [Pg 245] of which two broad types may be recognised; first, those where the pieces are dealt with in the form of ropes in a twisted form, and, second, those where the pieces are washed open. There are some machines in which the cloths may be treated either in the open or rope form as may be thought most desirable.

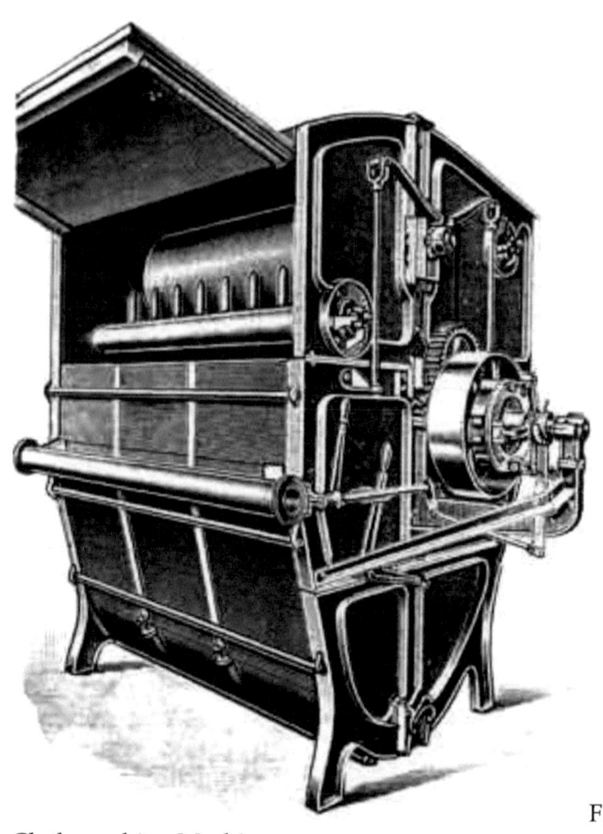

FIG. 34. —
Cloth-washing Machine.

[Pg 246]

Fig. 33 represents a fairly well-known machine, made by Messrs. Mather & Platt, in which the pieces are treated in a rope-like form. It consists of a trough in which a constant current of water is maintained. At one end of this trough is a square beating roller, at the other a wooden lattice roller. Above the square beater, and out of the trough, are a pair of rollers whose purpose is to draw the cloth through the machine and also partly to act as squeezing rollers. As will be seen the cloth is threaded in rope form spirally round the rollers, passing in at one end and out at the other, pegs in a guide rail serving to keep the various portions separate. The square beater in its revolutions has a beating action on the cloth tending to more effectual washing. The lattice roller is simply a guide roller.

Fig. 34 shows a washing machine very largely used in the wool-dyeing trade. The principal portion of this machine is of wood.

The internal parts consist of a large wooden bowl, or oftener, as in the machine under notice, of a pair of wooden bowls which are pressed together by springs with some small degree of force. Between these bowls the cloth is placed, more or less loosely twisted up in a rope form, and the machines are made to take four, six or eight pieces, or lengths of pieces, at one time, the ends of the pieces being stitched together. A pipe running along the front of the machine conveys a constant current of clean water which is caused to impinge in the form of jets on the pieces of cloth as they run through the machine, while an overflow carries away the used water. The goods are run in this machine until they are considered to be sufficiently washed, which may take half to one and a half hours.

In Fig. 35 is shown a machine designed to wash pieces in the broad or open state. The machine contains a large number of guide rollers, built more or less open, round which [Pg 247] the pieces are guided — the ends of the pieces being stitched together. Pipes carrying water are so arranged that jets of clean water impinge on and thoroughly wash the cloth as it passes through — the construction of the guide rollers facilitating the efficient washing of the goods.

FIG. 35.—Cloth-washing Machine.

Soaping.—- Sometimes yarns or cloths have to be passed through a soap bath after being dyed in order to brighten up the colours or develop them in some way. In the case of yarns this can be done on the reel washing machine such as is shown in Fig. 32. In the case of piece goods, a continuous machine, in which the washing, soaping, etc., can be carried on simultaneously, is often employed. Such a machine is shown in Fig. 36. It consists of a number of compartments fitted with guide rollers so that the cloth passes up and down several times through the liquors in the [Pg 248] compartments. Between one compartment and another is placed a pair of squeezing rollers. The cloth is threaded in a continuous manner, well shown in the drawing, through the machine. In one compartment it is treated with water, in another soap liquor, in another water, and so on; and these machines may be made with two, three or more compartments as may be necessary for the particular work in hand. As seen in the drawing, the cloth passes in at one end and out at the other finished. It is usually arranged that a continuous current of the various liquors used flows through the various compartments, thus ensuring the most perfect treatment of the cloths.

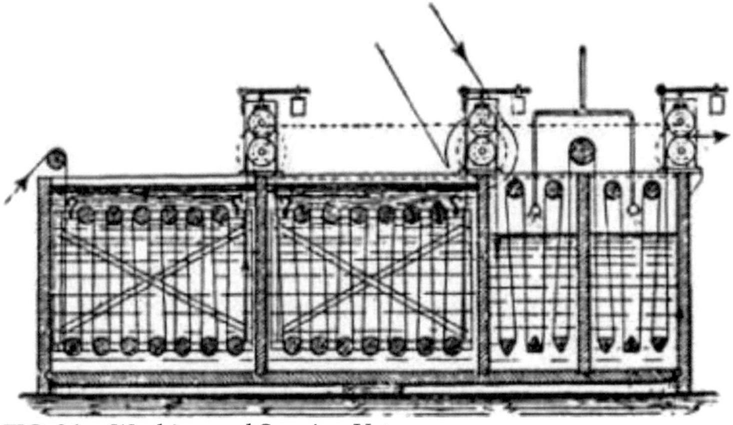

FIG. 36.—Washing and Soaping Vats.

Steaming.—Sometimes it becomes necessary to subject dyed goods to a process of steaming, as, for instance, with steam aniline blacks, khaki shades, alizarine reds, etc., for the purpose of more fully developing and fixing the dye upon the fibre. In the case of yarns, this operation is carried out in the steaming cottage, one form of which is shown in Fig. 37. It consists of a horizontal cylindrical iron vessel like a steam boiler, one end is entirely closed, while the other is made to open and be closed tightly and hermeti [Pg 249] cally. The cottage is fitted with the necessary steam inlet and outlet pipes, drain pipes for condensed water, pressure gauges. The yarn to be steamed is hung on rods placed on a skeleton frame waggon on wheels which can be run in and out of the steaming cottage as is required. The drawing shows well the various important parts of the machine. In the case of piece goods these also can be hung from rods in folds on such a waggon, but it is much more customary to employ a continuous steaming chamber, very similar to

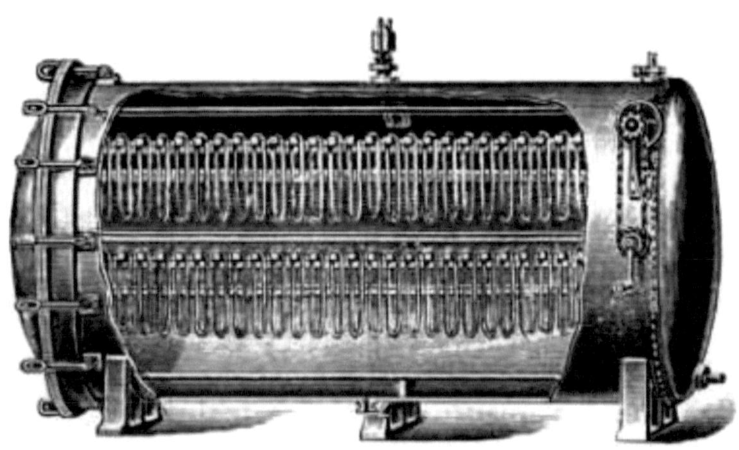

FIG. 37.—Steaming Cottage.

the ageing and oxidising machine shown in Fig. 38, and also used in the dyeing of aniline black.

Drying.—Following on the washing comes the final operation of the dyeing process, that of drying the dyed and washed goods. Textile fabrics of all kinds after they have passed through dye-baths, washing machines, etc., contain a large amount of water, often exceeding in weight that of the fibre itself, and to take the goods direct from the preceding operations to the drying plant means that a considerable amount of fuel must be expended to drive [Pg 250]

FIG. 38.—Steaming and Ageing Chamber.

[Pg 251]

off this large amount of water. It is therefore very desirable that the goods be freed from as much of this water as possible before they are sent into any drying chambers, and this may be done in three ways, by wringing, squeezing and hydro-extracting. The first two methods have already been described (p. 239, etc.) and need not again be alluded to; the last needs some account.

FIG. 39.—Hydro-extractor.

Hydro-extractors are a most efficient means for extracting water out of textile fabrics. They are made in a variety of forms by several makers. Essentially they consist of a cylindrical vessel, or basket, as it is called, with perforated sides so constructed that it can be revolved at a high speed. This vessel is enclosed in an outer cage. The goods are placed in the basket, as it is termed, and then this is caused to revolve at high speed, when centrifugal action comes into play, and the water contained in the goods finds its way to the outside of the basket through the perforations, [Pg 252] and so away from the goods. Hydro-extractors are made in a variety of sizes and forms—in some the driving gear is above, in others below the basket; in some the driving is done by belt-gearing, in others a steam engine is directly connected with the basket. Figs. 39 and 40 show two forms which are much in use in the textile industry. They are very efficient, and extract water from textile goods more completely than any other means, as will be obvious from a study of the table below.

FIG. 40.—Hydro-extractor.

The relative efficiency of the three systems of extracting the moisture out of textile fabrics has been investigated by Grothe, who gives in his *Appretur der Gewebe*, published in 1882, the following table showing the percentage amount of water removed in fifteen minutes:—

Yarns. Wool. Silk. Cotton. Linen. Wringing 44.5 45.4 45.3 50.3 Squeezing 60.0 71.4 60.0 73.6 Hydro-extracting 83.5 77.0 81.2 82.8

Pieces. Wringing 33.4 44.5 44.5 54.6 Squeezing 64.0 69.7 72.2 83.0 Hydro-extracting 77.8 75.5 82.3 86.0 [Pg 253]

In the practical working of hydro-extractors it is of the utmost importance that the goods be carefully and regularly laid in the basket—not too much in one part and too little in another. Any unevenness in this respect at the speed at which they are driven lays such a strain on the bearings as to seriously endanger the safety of the machine.

After being wrung, squeezed or hydro-extracted the goods are ready to be dried. In the case of yarns, this may be done in rooms heated by steam pipes placed on the floor, the hanks being hung on rods suspended from racks arranged for the purpose.

FIG. 41.—Automatic Yarn-dryer.

Where large quantities of yarn have to be dried, it is most economical to employ a yarn-drying machine, and one form of such is shown in Fig. 41. The appearance of the machine is that of one long room from the outside; internally it is divided into compartments, each of which is heated up by suitably arranged steam pipes, but the degree of heating in each compartment varies—at the entrance end it is high, at the exit end lower. The yarn is fed in at one end, being hung on rods, and by suitable gearing it is carried directly through the various chambers or sections, and in its passage the heat to which it is subjected drives off the water it contains. The yarn requires no attention [Pg 254] from the time it passes in wet at the one end of the machine and comes out dry at the other end. The amount of labour required is slight, only that represented by filling the sticks with wet yarn and emptying them of the dried yarn. The machine works regularly and well.

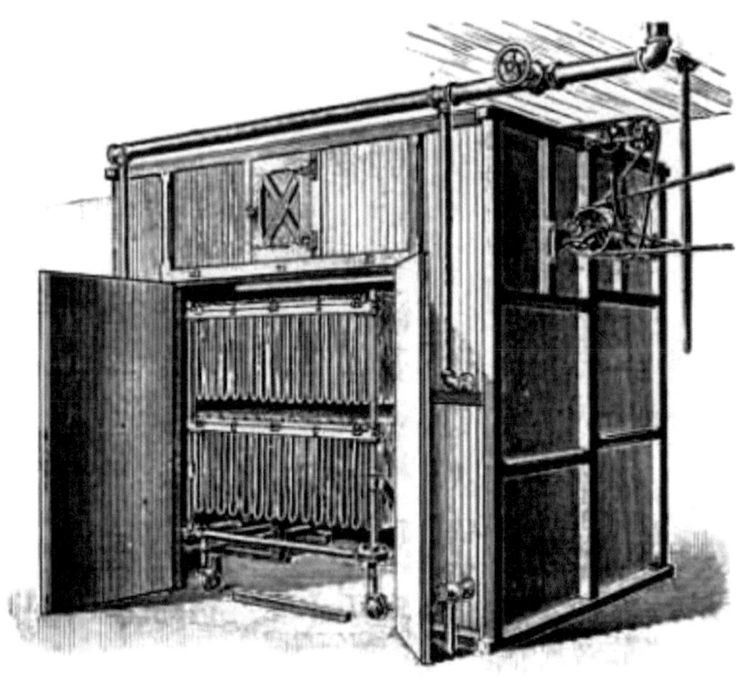

FIG. 42. — Truck Yarn-dryer.

The drying is accomplished by circulating heated air through the yarns, this heating being effected by steam coils, fresh air continually enters the chambers, while water-saturated air is as continually being taken out at the top of the chamber. One of the great secrets in all drying operations is to have a constant current of fresh hot air playing on the goods to be dried; this absorbs the moisture [Pg 255] they contain, and the water-charged air thus produced must be taken away as quickly as possible.

FIG 43.—Drying Cylinders.

Fig. 42 shows what is called a truck yarn-dryer, which consists of a chamber heated with steam pipes and fitted with an exhausting fan to draw out the air and water vapour which is produced. The yarns are hung on trucks [Pg 256] which can be run in and out of the chamber for filling and emptying.

Piece Goods.—The most convenient manner of drying piece goods is to employ the steam cylinder drying machine, such as is shown in Fig. 43. This consists of a number of hollow tin or copper cylinders which can be heated by steam passing in through the axles of the cylinders, which are made hollow on purpose. The cloth to be dried passes round these cylinders, which revolve while the cloth passes. They work very effectually. The cylinders are arranged sometimes, as in the drawing, vertically; at other times horizontally. [Pg 257]

CHAPTER VIII

TESTING OF THE COLOUR OF DYED FABRICS.

It is frequently desirable that dyers should be able to ascertain with some degree of accuracy what dyes have been used to dye any particular samples of dyed cloth that has been offered to them to match. In these days of the thousand and one different dyes that are known it is by no means an easy thing to do; and when, as is most often the case, two or three dye-stuffs have been used in the production of a shade, the difficulty is materially increased.

The only available method is to try the effect of various acid and alkaline reagents on the sample, noting whether any change of colour occurs, and judging accordingly. It would be a good thing for dyers to accustom themselves to test the dyeings they do, and so accumulate a fund of practical experience which will stand them in good stead whenever they have occasion to examine a dyed pattern of unknown origin.

The limits of this book does not permit of there being given a series of elaborate tables showing the action of various chemical reagents on fabrics dyed with various colours; and such, indeed, serve very little purpose, for it is most difficult to describe the minor differences which often serve to distinguish one colour from another. Instead of doing so, we will point out in some detail the methods of carrying out the various tests, and advise all dyers to carry these out for themselves on samples dyed with known [Pg 258] colours, and when they have an unknown colour to test to make tests comparatively with known colours that they think are likely to have been used in the production of the dyed fabric they are testing.

One very common method is to spot the fabric, that is, to put a drop of the reagent on it, usually with the end of the stopper of the reagent bottle, and to observe the colour changes, if any, which ensue. This is a very useful test and should not be omitted, and it is often employed in the testing of indigo dyed goods with nitric acid, those of logwood with hydrochloric acid, alizarine with caustic soda, and many others. It is simple and easy to carry out, and only takes a few minutes.

To make a complete series of tests of dyed fabrics there should be provided the following reagents:—

1. Strong sulphuric acid, as bought.

2. Dilute sulphuric acid, being the strong acid diluted with twenty times its volume of water.

3. Concentrated hydrochloric acid.

4. Dilute hydrochloric acid, 1 acid to 20 water.

5. Concentrated nitric acid.

6. Dilute nitric acid, 1 acid to 20 water.

7. Acetic acid.

8. Caustic soda solution, 5 grams in 100 c.c. water.

9. Ammonia (strong).

10. Dilute ammonia, 1 strong ammonia to 10 water.

11. Carbonate of soda solution, 6 grams in 100 c.c. water.

12. Bleaching powder solution, 2° Tw.

13. Bisulphite of soda, 72° Tw.

14. Stannous chloride, 10 grams crystals in 100 c.c. water, with a little hydrochloric acid.

15. Methylated spirit.

Small swatches of the dyed goods are put in clean porcelain basins, and some of these solutions poured over them. Any [Pg 259] change of colour of the cloth is noted, as well as whether any colour is imparted to the solutions. After making observations of the effects in the cold the liquids may be warmed and the results again noted. After being treated with the acids the swatches should be well washed with water, when the original colour may be wholly or partially restored.

To give tables showing the effects of these reagents on the numerous dyes now known would take up too much room and not serve a very useful purpose, as such tables, if too much relied on, leave the operator somewhat uncertain as to what he has before him. The reader will find in Hurst's *Dictionary of Coal-tar Colours* some useful notes as to the action of acids and alkalies on the various colours that may be of service to him.

Alizarine and the series of dye-stuffs to which it has given its name, fustic, cochineal, logwood and other dyes of a similar class require the fabric to be mordanted, and the presence of such mordant is occasionally an indirect proof of the presence of these dyes.

To detect these mordants, a piece of the swatch should be burnt in a porcelain or platinum crucible over a Bunsen burner, care being taken that all carbonaceous matter be burnt off. A white ash will indicate the presence of alumina mordants, red ash that of iron mordants, and a greenish ash chrome mordants.

To confirm these the following chemical tests may be applied: Boil the ash left in the crucible with a little strong hydrochloric acid and dilute with water. Pass a current of sulphuretted hydrogen gas through the solution; if there be any tin present a brown precipitate of tin sulphide will be obtained. This can be filtered off. The filtrate is boiled for a short time with nitric acid, and ammonia is added to the solution when alumina is thrown down as a white gelatinous precipitate; iron is thrown down as a brown red bulky pre [Pg 260] cipitate; while chrome is thrown down as a greyish-looking gelatinous precipitate. The precipitate obtained with the ammonia is filtered off, and a drop of ammonium sulphide added, when any zinc present will be thrown down as white precipitate of zinc sulphide: to the filtrate from this ammonium oxalate may be added, when if lime is present a white precipitate of calcium oxalate is obtained.

A test for iron is to dissolve some of the ash in a little hydrochloric acid, and add a few drops of potassium ferrocyanide solution, when if any iron be present a blue precipitate will be obtained.

To make more certain of the presence of chrome, heat a little of the ash of the cloth with caustic soda and chlorate of soda in a porcelain crucible until well fused, then dissolve in water, acidify with acetic acid and add lead acetate; a yellow precipitate indicates the presence of chrome.

A book on qualitative chemical analysis should be referred to for further details and tests for metallic mordants.

The fastness of colours to light, air, rubbing, washing, soaping, acids and alkalies is a feature of some considerable importance. There are indeed few colours that will resist all these influences, and such are fully entitled to be called fast. The decree of fastness varies very considerably. Some colours will resist acids and alkalies well, but are not fast to light and air; some will resist washing and soaping, but are not fast to acids; Some may be fast to light, but are not so to washing. The following notes will show how to test these features:—

Fastness to Light and Air.—This is simply tested by hanging a piece of the dyed cloth in the air, keeping a piece in a drawer to refer to, so that the influence on the original colour can be noted from time to time. If the piece is left out in the open one gets not only the effect of light but also that of climate on the colour, and there is no doubt wind, rain, hail and snow have some influence on the fading of the colour. [Pg 261]

If the piece is exposed under glass, the climatic influences do not come into play, and one gets the effect of light alone.

In making tests of fastness, the dyer will and does pay due regard to the character of the influences that the material will be subjected to in actual use, and these vary very considerably; thus the colour of underclothing need not be fast to light, for it is rarely subjected to that agent of destruction. On the other hand it must be fast to washing, for that is an operation to which underclothing is subjected week by week.

Window curtains are much exposed to light and air, and, therefore, colours in which they are dyed should be fast to light and air. On the other hand these curtains are rarely washed, and so the colour need not be quite fast to washing. And so with other kinds of fabrics, there are scarcely two kinds which are subjected to the same influences, and require the colours to have the same degree of fastness.

The fastness to rubbing is generally tested by rubbing the dyed cloth on a piece of white paper.

Fastness to Washing.—This is generally tested by boiling a swatch of the cloth in a solution of soap containing 4 grams of a good neutral curd soap per litre for ten minutes and noting the effect—whether the soap solution becomes coloured and to what degree, or whether it remains colourless, and also whether the colour of the swatch has changed at all.

One very important point in connection with the soaping test is whether a colour will run into a white fabric that may be soaped along with it. This is tested by twisting strands of the dyed yarn or cloth with white yarn or cloth and boiling them in the soap liquor for ten minutes and then noting the effect, particularly observing whether the white pieces have taken up any colour.

Fastness to acids and fastness to alkalies is observed while carrying out the various acid and alkali tests given above. [Pg 262]

CHAPTER IX.

EXPERIMENTAL DYEING AND COMPARATIVE DYE TESTING.

Every dyer ought to be able to make experiments in the mordanting and dyeing of textile fibres for the purpose of ascertaining the best methods of applying mordants or dye-stuffs, the best methods of obtaining any desired shade, and for the purpose of making comparative tests of dyes or mordanting materials with the object of determining their strength and value. This is not by any means difficult, nor does it involve the use of any expensive apparatus, so that a dyer need not hesitate to set up a small dyeing laboratory for fear of the expense which it might entail.

In order to carry out the work indicated above there will be required several pieces of apparatus. First, a small chemical balance, one that will carry 100 grams in each pan is quite large enough; and such a one, quite accurate enough for this work, can be bought for 25s. to 30s., while if the dyer be too poor even for this, a cheap pair of apothecaries' scales might be used. It is advisable to procure a set of gram weights, and to get accustomed to them, which is not a very difficult task.

In using the balance always put the substance to be weighed on the left-hand pan, and the weights on the right-hand pan. Never put chemicals of any kind direct on the pan, but weigh them in a watch glass, small porcelain basin, or glass beaker, which has first been weighed, [Pg 263] according to the nature of the material which is being weighed. The sets of weights are always fitted into a block or box, and every time they are used they should be put back into their proper place.

The experimenter will find it convenient to provide himself with a few small porcelain basins, glass beakers, cubic centimetre measures, two or three 200 c.c. flasks with a mark on the neck, a few pipettes of various sizes, 10 c.c., 20 c.c., 25 c.c.

The most important feature is the dyeing apparatus. Where only a single dye test is to be made, a small copper or enamelled iron saucepan, such as can be bought at any ironmonger's, may be used; this may conveniently be heated by a gas boiling burner, such as can also be bought at an ironmonger's or plumber's for 2s.

FIG. 44. — Experimental Dye-bath.

It is, however, advisable to have means whereby several dyeing experiments can be made at one time and under precisely the same conditions, and this cannot be done by using the simple means noted above.

To be able to make perfectly comparative dyeing experiments it is best to use porcelain dye-pots — these may be bought from most dealers in chemical apparatus — and to heat them in a water-bath arrangement.

The simplest arrangement is sketched in Fig. 44; it consists of a copper bath measuring 15 inches long by 10½ [Pg 264] inches broad and 6½ inches deep — this is covered by a lid in which are six apertures to take the porcelain dye-baths. The bath is heated by two round gas boiling burners of the type already referred to.

The copper bath is filled with water, which, on being heated to the boil by the gas burners, heats up the dye liquids in the dye-pots. The temperature in the dye-pots under such conditions can never reach the boiling point; where it is desirable, as in some cases of wool mordanting and dyeing, that it should boil, there should be added to the water in the copper bath a quantity of calcium chloride, which forms a solution that has a much higher boiling point than that of water, and so the dye liquors in the dye-pots may be heated up to the boil.

An objection might be raised that with such an apparatus the temperature in every part of the bath may not be uniform, and so the temperature of the dye-liquors in the pots may vary also, and differences of temperature often have a considerable influence on the shade of the colour which is being dyed. This is a minor objection, which is more academic in its origin than of practical importance. To obviate this Mr. William Marshall of the Rochdale Technical School has devised a circular form of dye-bath, in which the temperature in every part can be kept quite uniform.

The dyeing laboratories of technical schools and colleges are generally provided with a more elaborate set of dyeing appliances. These, in the latest constructed, consist of a copper bath supported on a hollow pair of trunnions, that the bath can be turned over if needed. Into the bath are firmly fixed three earthenware or porcelain dye-pots; steam for heating can be sent through the trunnions. After the dyeing tests have been made the apparatus can be turned over, and the contents of the dye-pots emptied into a sink which is provided for the purpose. [Pg 265]

Many other pieces of apparatus have been devised and made for the purpose of carrying on dyeing experiments on the small scale, but it will not be needful to describe these in detail. After all no more efficient apparatus can be desired than that described above.

Dyeing experiments can be made with either yarns or pieces of cloth—swatches, as they are commonly called—a very convenient size is a small skein of yarn or a piece of cloth having a weight of 5 grams. These test skeins or pieces ought to be well washed in hot water before use, so that they are clean and free from any size or grease. A little soda or soap will facilitate the cleansing process.

In carrying out a dyeing test the dye-pot should be filled with the water required, using as little as can be consistent with the dye swatch being handled comfortably therein, then there is added the required mordants, chemicals, dyes, etc., according to the character of the work which is being done.

Of such chemicals as soda, caustic soda, sodium sulphate (Glauber's salt), tartar, bichromate of potash, it will be found convenient to prepare stock solutions of known strength, say 50 grams per litre, and then by means of a pipette any required quantity can be conveniently added. The same plan might be followed in the case of dyes which are constantly in use, in this case, 5 grams per litre will be found strong enough.

Supposing it is desired to make a test of a sample of direct red, using the following proportions: 2 per cent. dye-stuff, 3 per cent. soda, 15 per cent. Glauber's salt, and the weight of the swatch which is being used is 5 grams. The following calculations are to be made to give the quantities of the ingredients required. [Pg 266]

For the dye-stuff: —

5 (weight of swatch) multiplied by 2 (per cent. of dye) and divided by 100 equals —

$$\frac{5 \times 2}{100} = 0.1 \text{ gram dye}$$

For the soda we have similarly: —

$$\frac{5 \times 3}{100} = 0.15 \text{ gram soda.}$$

For the Glauber's salt: —

$$\frac{5 \times 15}{100} = 0.75 \text{ gram Glauber's salt.}$$

These quantities may be weighed out and added to the dye-bath, or if solutions are kept, a calculation can be made as to the number of cubic centimetres which contain the above quantities, and these measured out and added to the dye-bath.

When all is ready, the bath is heated up, the swatch entered, and the work of the test entered upon.

Students are recommended to make experiments on such points as:

The shades obtained by using various proportions of dye-stuffs.

The influence of various assistants—common salt, soda, Glauber's salt, borax, phosphate of soda—in the bath.

The influence of varying proportions of mordants on the shade of dyeing.

The value of various assistants, tartar, oxalic acid, lactic acid, sulphuric acid, on the fixation of mordants.

The relative value of different tannin matters, etc.

Each dyer should make himself a pattern-book into which he should enter his tests, with full particulars as to how they have been produced at the side.

It is important that a dyer should be able to make [Pg 267] comparative dye tests to ascertain the relative strength of any two, or more samples of dyes which may be sent to him. This is not difficult, but requires considerable care in carrying out the various operations involved.

Of each of the samples of dyes 0.5 gram should be weighed out and dissolved in 100 c.c. of water, care being taken that every portion of the dye is dissolved before any of the solution is used in making up the dye-vats. Care should also be taken that the skeins of yarn or swatches of cloth are exactly equal in weight; that the same volume of water is placed in each of the dye-pots; that the same amount of sulphate of soda or other dye assistants are added; that the quantities of dye-stuff and solutions used are equal; in fact, that in all respects the conditions of dyeing are exactly the same, such, in fact, being the vital conditions in making comparative dye tests of the actual dyeing strength of several samples of dyes.

After the swatches have been dyed they are rinsed and then dried, when the depth of shade dyed on them may be compared one with another. To prevent any mistakes, it is well to mark the swatches with one, two, three or more cuts as may be required.

It is easier to ascertain if two dyes are different in strength of colour than to ascertain the relative difference between them. There are two plans available for this purpose—one is a dyeing test, the other is a colorimetric test made with the solutions of the dyes.

Dyeing Test.—This method of ascertaining the relative value of two dyes as regards strength of colour is carried out as follows: A preliminary test will show which is stronger than the other. Then there is prepared a series of dye-vats—one contains a swatch with the deepest of the two dyes, which is taken as the standard; the others, swatches with the other dye, but containing 2, 5 and 10 per cent. more dye [Pg 268] stuff, and all these swatches are dyed together, and after drying a comparison can be made between them and the standard swatch and a judgment formed as to the relative strength of the two dyes. A little experience will soon enable the dyer to form a correct judgment of the difference in strength between two samples of dye-stuff.

Colorimetric Test.—This is based on the principle that the colour of a solution of dye-stuff will be proportionate to its strength. Two white glass tubes equal in diameter are taken. Solutions of the dye-stuff, 0.5 gram in 100 c.c. of water, are prepared, care being taken that the solution is complete. Of one of these solutions 5 c.c. is taken and placed in one of the glass tubes, and 5 c.c. of the other solution is placed in the other glass tube. Of water 25 c.c. is now added to each tube, and then the colour of the diluted liquids is compared by looking through them in a good light. That sample which gives the deeper solution is the stronger in colouring power. By diluting the stronger solution with water until it is of the same depth of colour as the weaker, it may be assumed that the depth of the columns of liquid in the two tubes is in proportion to the relative strength of the two samples. Thus, if in one tube there are 30 c. of liquid and in the other 25 c., then the relative strength is as 30 to 25; and if the first is taken as the standard at 100, a proportion sum gives

30 : 25 : : 100 : 83.3,

that is, the weaker sample has only 83.3 per cent, of the strength of the stronger sample. [Pg 269]